Y0-BTA-816

Gerd Fischer
Jens Piontkowski

Ruled Varieties

Advanced Lectures in Mathematics

Editorial board:
Prof. Dr. Martin Aigner, Freie Universität Berlin, Germany
Prof. Dr. Michael Grüter, Universität des Saarlandes, Saarbrücken, Germany
Prof. Dr. Rudolf Scharlau, Universität Dortmund, Germany
Prof. Dr. Gisbert Wüstholz, ETH Zürich, Switzerland

Introduction to Markov Chains
Erhard Behrends

Einführung in die Symplektische Geometrie
Rolf Berndt

Wavelets – Eine Einführung
Christian Blatter

Local Analytic Geometry
Theo de Jong, Gerhard Pfister

Ruled Varieties
Gerd Fischer, Jens Piontkowski

Dirac-Operatoren in der Riemannschen Geometrie
Thomas Friedrich

Hypergeometric Summation
Wolfram Koepf

The Steiner Tree Problem
Hans-Jürgen Prömel, Angelika Steger

The Basic Theory of Power Series
Jésus M. Ruiz

vieweg

Gerd Fischer
Jens Piontkowski

Ruled Varieties

**An Introduction to
Algebraic Differential Geometry**

vieweg

Prof. Dr. Gerd Fischer
Dr. Jens Piontkowski
Heinrich-Heine-Universität Düsseldorf
Mathematisches Institut
Universitätsstraße 1
40255 Düsseldorf, Germany

gerdfischer@cs.uni-duesseldorf.de
piontkow@cs.uni-duesseldorf.de

Die Deutsche Bibliothek – CIP-Cataloguing-in-Publication-Data
A catalogue record for this publication is available from
Die Deutsche Bibliothek.

First edition, May 2001

All rights reserved
© Friedr. Vieweg & Sohn Verlagsgesellschaft mbH, Braunschweig/Wiesbaden, 2001

Vieweg is a company in the specialist publishing group BertelsmannSpringer.

No part of this publication may be reproduced, stored in a retrieval system or
transmitted, mechanical, photocopying or otherwise without prior permission
of the copyright holder.

www.vieweg.de

Cover design: Ulrike Weigel, www.CorporateDesignGroup.de
Printing and binding: Hubert & Co, Göttingen
Printed on acid-free paper
Printed in Germany

ISBN 3-528-03138-7   ISSN 0932-7134

Er redet nur von Regelflächen,
ich werd' mich an dem Flegel rächen!

Dedicated to the memory of Karl Stein (1913 - 2000)

Preface

The simplest surfaces, aside from planes, are the traces of a line moving in ambient space or, more precisely, the unions of one-parameter families of lines. The fact that these lines can be produced using a ruler explains their name, "ruled surfaces." The mechanical production of ruled surfaces is relatively easy, and they can be visualized by means of wire models. These surfaces are not only of practical use, but also provide artistic inspiration.

Mathematically, ruled surfaces are the subject of several branches of geometry, especially differential geometry and algebraic geometry. In classical geometry, we know that surfaces of vanishing Gaussian curvature have a ruling that is even developable. Analytically, developable means that the tangent plane is the same for all points of the ruling line, which is equivalent to saying that the surface can be covered by pieces of paper. A classical result from algebraic geometry states that rulings are very rare for complex algebraic surfaces in three-space: Quadrics have two rulings, smooth cubics contain precisely twenty-seven lines, and in general, a surface of degree at least four contains no line at all. There are exceptions, such as cones or tangent surfaces of curves. It is also well-known that these two kinds of surfaces are the only developable ruled algebraic surfaces in projective three-space.

The natural generalization of a ruled surface is a ruled variety, i.e., a variety of arbitrary dimension that is "swept out" by a moving linear subspace of ambient space. It should be noted that a ruling is not an intrinsic but an extrinsic property of a variety, which only makes sense relative to an ambient affine or projective space. In this book, we consider ruled varieties mainly from the point of view of complex projective algebraic geometry, where the strongest tools are available. Some local techniques could be generalized to complex analytic varieties, but in the real analytic or even differentiable case there is little hope for generalization: the reason being that rulings, and especially developable rulings, have the tendency to produce severe singularities.

As in the classical case of surfaces, there is a strong relationship between the subject of this book, ruled varieties, and differential geometry. For our purpose, however, the Hermitian Fubini-Study metric and the related concepts of curvature are not necessary. In order to detect developable rulings, it suffices to consider a bilinear second fundamental form that is the differential of the Gauß map. This method does not give curvature as a number, but rather measures the degree of vanishing of curvature; this point of view has been used in a fundamental paper of GRIFFITHS and HARRIS [GH₂]. The purposes of this book are to make parts of this paper more accessible, to give detailed and more elementary proofs, and to report on recent progress in this area. This text can also serve as an introduction to "bilinear" complex differential geometry, a useful method in algebraic geometry.

In Chapter 0 we recall some classical facts about developable surfaces in real affine and complex projective three-space. The basic results are *a)* the linearity of the fibers of the Gauß map and *b)* the classification of developable surfaces, locally in the real differentiable case and globally in the complex projective case. We present very elementary proofs, which can be adapted to more general situations.

Families of linear spaces correspond to subsets of Grassmannians. Chapter 1 contains all the facts we need about these varieties and more; for instance, there is a meticulous and yet simple proof of the Plücker relations, which is not easy to find in the existing literature. Varieties that are unions of one-parameter families of linear spaces correspond to curves in the Grassmannian. Following [GH₂], we derive a normal form representation of such curves, which then implies the classification of developable varieties of Gauß rank one in Chapter 2.

Chapter 2 is the core of the book. We first introduce the different concepts of rulings of a variety and methods to construct them. Then we concentrate on the additional condition of developability; this class of ruled varieties is accessible to the methods of differential geometry, especially the Gauß map and the associated second fundamental form. The basic theorem of this chapter is the linearity of the fibers of the Gauß map, for which we give proofs from various points of view. This result shows the Gauß ruling to be the maximal developable ruling and the Gauß rank to be an invariant of the variety.

The central problem concerning developable varieties is their classification, and there are two cases where it can be solved:

  – Arbitrary dimension and Gauß rank one (2.2.8).

  – Codimension one and Gauß rank two (2.5.6).

The key to the second case is duality or, more precisely, the dual variety of a given variety, which is a subvariety of the dual projective space.

Chapter 3 is devoted to a very important class of ruled varieties: tangent and secant varieties. In general, the canonical ruling of the tangent variety is not developable, but it contains a developable subruling by lines. An amazing result is the formation of a developable ruling with larger fiber dimension if the tangent variety has a smaller than expected dimension. The dimension of the tangent variety, like nearly all other invariants in this book, can be computed from the second fundamental form. This is not the case, however, for the dimension of the secant variety, where third order invariants become important. Until now, all attempts to find the right third or higher order invariant for the problem of this dimension have failed, even for smooth varieties. Thus the book closes with one more open problem.

The only prerequisites for this textbook are the basic concepts of complex affine and projective algebraic geometry, the results of which are outlined in Chapter 1. The relatively elementary methods that we use throughout are based on linear algebra with algebraic or holomorphic parameters.

We wish to express our thanks to our collaborators Hung-Hsi Wu and Joseph Landsberg, to Gabriele Süß for producing the camera-ready TEX-manuscript, and to our students for their help in proofreading. We are also indebted to Vieweg for publishing this advanced text.

Düsseldorf, March 2001   Gerd Fischer

                                                                                       Jens Piontkowski

Contents

Chapter 0

Review from Classical Differential and Projective Geometry

In this introductory chapter we recall some fairly well known facts about surfaces in "three-space". The classical literature does not make careful distinctions between the different kinds of three-spaces. Nowadays, we are urged to be more precise: we start in the real affine space $\mathbb{R}^3$ and end up in complex projective space $\mathbb{P}_3(\mathbb{C})$. The reader will notice that we try to avoid all kinds of parameterizations by arc length and other typical C^∞-arguments since they cannot be generalized to the holomorphic case.

0.1 Developable Rulings

We start with a *surface* $X \subset \mathbb{R}^3$, which is assumed to be sufficiently differentiable. This surface is not necessarily closed since our investigations are only local. Furthermore, it is not adequate to assume that X is smooth everywhere, since it will turn out that certain singularities are unavoidable.

If $p \in X$ is a smooth point, there is a tangent plane $T_p X \subset \mathbb{R}^3$, which we consider as a vector subspace. The set of two-dimensional vector subspaces of $\mathbb{R}^3$ is $\mathbb{P}_2^*(\mathbb{R})$, the dual of the real projective plane $\mathbb{P}_2(\mathbb{R})$. Hence, if X is smooth, we have a *Gauß map*

$$\gamma : X \longrightarrow \mathbb{P}_2^*(\mathbb{R}), \quad p \longmapsto T_p X.$$

If X is orientable (for local questions this is no restriction), there is a "lifted" *Gauß map*

$$\mathfrak{n} : X \longrightarrow S_2 \subset \mathbb{R}^3, \quad p \longmapsto \mathfrak{n}(p),$$

where S_2 is the two-dimensional sphere with radius 1 and $\mathfrak{n}(p)$ is the unit normal vector of $T_p X$.

For every smooth point $p \in X$, there are open neighborhoods $p \in U \subset X$ and $V \subset \mathbb{R}^2$ with a *parameterization* (i.e. a diffeomorphism)

$$\varphi : V \longrightarrow U.$$

We say that X is *ruled around* p if there is a parameterization of a special form: the neighborhood V must be of the type

$$V = I \times J \quad \text{with } I \subset \mathbb{R} \quad \text{and } J \subset \mathbb{R}$$

open intervals containing the origin o, and there must exist curves

$$\alpha : I \longrightarrow U \quad \text{and} \quad \varrho : I \longrightarrow \mathbb{R}^3 \setminus \{0\}$$

such that $\varphi(o, o) = p$ and

$$\varphi(s, t) = \alpha(s) + t\varrho(s) \quad \text{for every } (s, t) \in I \times J \, .$$

The curve α is called a *directrix* and the segments of lines $\varphi(s, J) \subset X$ are called *generators* of the ruling.

It should be noted that the concept of a ruling is no intrinsic property of the surface. It depends on the ambient space $\mathbb{R}^3$, where linearity is measured.

The ruling of X around p defined above is called *developable* if the tangent plane of X is the same for all points of a generator $\varphi(s, J)$, i.e. the Gauß map is constant along the generators. The tangent plane of X is generated by the columns of the Jacobian

$$\text{Jac } \varphi = \left(\alpha' + t\varrho', \varrho \right) , \quad \text{where } ' = \frac{d}{ds} \, .$$

Hence it is evident that developability is equivalent to $\varrho' \in \text{span} \{\alpha', \varrho\}$ or

$$\det \left(\alpha', \varrho, \varrho' \right) = 0 \quad \text{on } I \, .$$

There are three obvious examples for developable rulings, given as the image of a map

$$\varphi : I \times \mathbb{R} \longrightarrow \mathbb{R}^3 , \quad (s, t) \longmapsto \alpha(s) + t\varrho(s) \, .$$

1. If ϱ is constant and α is a curve such that $\alpha'(s) \notin \mathbb{R} \varrho$ for every $s \in I$, then φ is an immersion, and $X = \varphi(I \times \mathbb{R})$ is a *cylinder*. Since $\varrho' = 0$, the local ruling is developable.

2. If α is constant, $\alpha(I) = p$, and ϱ has the property $\varrho'(s) \notin \mathbb{R}\varrho(s)$, then φ is an immersion for $t \neq 0$ and $X = \varphi(I \times \mathbb{R})$ is a *cone* with vertex p. Since $\alpha' = 0$, the local ruling is developable.

3. If α is a curve, such that $\alpha'(s)$ and $\alpha''(s)$ are linearly independent for every $s \in I$ (they generate the osculating plane), we may define $\varrho := \alpha'$. Then φ is an immersion for $t \neq 0$, and $X = \varphi(I \times \mathbb{R})$ is the *tangent surface* of the curve α; again the local ruling is developable.

In all the three cases, X is the image of an immersion; thus, we cannot expect any global statements. Certainly, in examples 2 and 3 one cannot avoid singularities if the generators are complete lines. The cone is singular in its vertex, and the tangent surface is singular along the distinguished directrix $\alpha(I)$.

In 0.4 we shall see that the above list of examples is complete from the local point of view.

0.2   Vanishing Gauß Curvature

Recall that a surface $X \subset \mathbb{R}^3$ has two *principal curvatures*

$$\kappa_1(p), \kappa_2(p) \in \mathbb{R}$$

in every smooth point p and that the *Gauß curvature* is

$$\kappa_G(p) := \kappa_1(p) \cdot \kappa_2(p).$$

The point $p \in X$ is called *planar* if $\kappa_1(p) = \kappa_2(p) = 0$. The principal curvatures are the eigenvalues of the second fundamental form, which is the differential of the Gauß map.

It is a classical result that surfaces of constant positive resp. negative Gauß curvature are spheres, resp. pseudospheres. Cases in which the Gauß curvature disappears everywhere are more difficult to handle. It is easy to see that a developable surface has Gauß curvature zero, e.g. [Kli, 3.7.5]. Locally, the converse is true:

Linearity Theorem. *Let $X \subset \mathbb{R}^3$ be a surface with a nonplanar point $p \in X$, such that the Gauß curvature κ_G vanishes in a neighborhood of p. Then X is ruled around p by the fibers of the Gauß map. In particular, the fibers of the Gauß map around p are segments of lines, and the ruling is developable.*

Proof. The assumptions on the curvature imply that the Jacobian of the Gauß map

$$\gamma : X \longrightarrow \mathbb{P}_2^*(\mathbb{R})$$

has rank one in a neighborhood of p. By the implicit function theorem we can find a parameterization

$$\varphi : I \times J \longrightarrow U \subset X, \quad (s, t) \longmapsto \varphi(s, t),$$

where U is a neighborhood of p and $I, J \subset \mathbb{R}$ are intervals containing the origin o, such that γ is constant on $\varphi(\{s\} \times J)$ for every $s \in I$.

Now we replace the Gauß map γ by a map

$$\widetilde{\gamma} : I \times J \longrightarrow \mathbb{R}^3 \setminus \{0\} \quad \text{with} \ \langle \widetilde{\gamma}(s, t), T_{\varphi(s,t)}X \rangle = 0$$

for all $(s, t) \in I \times J$. Denoting derivatives by subscripts, let us write

$$\langle \widetilde{\gamma}, \varphi_s \rangle = 0, \quad \text{and} \tag{1}$$

$$\langle \widetilde{\gamma}, \varphi_t \rangle = 0. \tag{2}$$

The assumption on the fibers of the Gauß map implies that there is a differentiable function $\lambda : I \times J \longrightarrow \mathbb{R}$ such that

$$\widetilde{\gamma}_t = \lambda \widetilde{\gamma} \, . \tag{3}$$

In order to show the linearity of $\varphi(\{s\} \times J)$ we must prove

$$\varphi_{tt} = \mu \varphi_t \tag{4}$$

for some differentiable function μ (this can be seen as in 1.1.5, where the complex analogue is proved).

First we show that for every (s, t), we have

$$\mathbb{C}\varphi_t = \{ v \in T_{\varphi(s,t)} X : \langle \widetilde{\gamma}_s, v \rangle = 0 \} \, . \tag{5}$$

Differentiating (1) with respect to t yields

$$\langle \widetilde{\gamma}_t, \varphi_s \rangle + \langle \widetilde{\gamma}, \varphi_{st} \rangle = 0 \, ;$$

together with (3) this implies $\langle \widetilde{\gamma}, \varphi_{st} \rangle = 0$. If we plug this into the derivative of (2) with respect to s we obtain

$$\langle \widetilde{\gamma}_s, \varphi_t \rangle = 0 \, , \tag{6}$$

which implies that the left hand side of (5) is contained in the right hand side. So both are the same, because p is a nonplanar point.

In order to prove (4) we have to show

$$\langle \widetilde{\gamma}, \varphi_{tt} \rangle = 0 \, , \quad \text{i.e. } \varphi_{tt} \in T_{\varphi(s,t)} X \, , \quad \text{and} \tag{7}$$

$$\langle \widetilde{\gamma}_s, \varphi_{tt} \rangle = 0 \, , \quad \text{i.e. } \varphi_{tt} \in \mathbb{C}\varphi_t \, . \tag{8}$$

Differentiating (2) with respect to t yields

$$\langle \widetilde{\gamma}_t, \varphi_t \rangle + \langle \widetilde{\gamma}, \varphi_{tt} \rangle = 0 \, ;$$

together with (3) we obtain (7). By differentiating (6) with respect to t, we finally have

$$\langle \widetilde{\gamma}_{ts}, \varphi_t \rangle + \langle \widetilde{\gamma}_s, \varphi_{tt} \rangle = 0 \, ,$$

and plugging in the differential of (3) with respect to s yields (8) by (6).   □

There are slightly shorter proofs of this theorem, using curvature coordinates [Kli, 3.7]. However the above arguments are much more elementary and show that the statement has very little to do with curvature: it is about the structure of the Gauß map in case its rank is not maximal. The same method also works in the complex analytic case and for arbitrary hypersurfaces [FW, Appendix].

0.3 Hessian Matrices

Developability is a condition for the change of tangents, hence it can be expressed by second derivatives. We will show more explicitly how the Hessian matrix occurs. C. F. GAUSS [G, Sec. 9] has already given a beautiful formula for the Gauß curvature κ_G of a surface X given by an equation. If

$$X = \{x \in \mathbb{R}^3 : f(x) = 0\}, \quad \text{then}$$

$$\kappa_G = \frac{-1}{(f_1^2 + f_2^2 + f_3^2)^2} \det \begin{pmatrix} 0 & f_1 & f_2 & f_3 \\ f_1 & f_{11} & f_{12} & f_{13} \\ f_2 & f_{21} & f_{22} & f_{23} \\ f_3 & f_{31} & f_{32} & f_{33} \end{pmatrix},$$

where subscripts mean partial derivatives (see also [D]). The matrix $\overline{H}$ occurring here is called the *extended Hesse matrix* . If we use the results of 0.2, then it is clear that developability is essentially equivalent to $\det \overline{H} = 0$. Here we give a direct proof of this result without explicit use of curvature. It should be noted that a projective version of the Hesse determinant in case of curves is an indicator of *flexes* [F$_2$, 4.5].

If the surface $X \subset \mathbb{R}^3$ is defined by a differentiable function f as above, then the Gauß map is easy to describe. If

$$g_p := \operatorname{grad}_p f = (f_1(p), f_2(p), f_3(p))$$

is the *gradient* of f at $p \in X$, then

$$\gamma : X \longrightarrow \mathbb{P}_2^*(\mathbb{R}), \quad x \longmapsto (f_1(p) : f_2(p) : f_3(p)) = \pi(g_p),$$

where $\pi : \mathbb{R}^3 \setminus \{o\} \to \mathbb{P}_2^*(\mathbb{R})$ denotes the canonical projection.

If $y \in \mathbb{P}_2^*(\mathbb{R})$ and $o \neq z \in \mathbb{R}^3$ is lying over y (i.e. $y = \pi(z)$), then the tangent plane to $\mathbb{P}_2^*(\mathbb{R})$ in y can be identified with

$$T_y \mathbb{P}_2^*(\mathbb{R}) = \mathbb{R}^3 / \mathbb{R}z ,$$

by using the surjective differential

$$d_z \pi : T_z \mathbb{R}^3 = \mathbb{R}^3 \longrightarrow T_y \mathbb{P}_2^*(\mathbb{R}) .$$

Now the differential of the lifting

$$g : X \longrightarrow \mathbb{R}^3 \setminus \{o\}, \quad p \longmapsto g_p ,$$

of γ is described by the *Hesse matrix*

$$H_p = \big(f_{ij}(p)\big)_{1 \leq i, j \leq 3}$$

of f at $p \in X$, i.e. for every tangent vector $v \in T_p X \subset \mathbb{R}^3$ we have

$$d_p g(v) = H_p \cdot v \in T_z \mathbb{R}^3, \quad \text{where } z = g_p.$$

Consequently we get

$$d_p \gamma(v) = H_p \cdot v + \mathbb{R}\, {}^t g_p \in T_{\gamma(p)} \mathbb{P}_2^*(\mathbb{R}) ;$$

this is the description of the second fundamental form we need. Its degeneracy is measured by

$$K_p := \operatorname{Ker} d_p \gamma = \{ v \in \mathbb{R}^3 : g_p \cdot v = 0,\ H_p \cdot v \in \mathbb{R}\, {}^t g_p \} \subset T_p X.$$

Now the *extended Hesse matrix*

$$\overline{H}_p := \begin{pmatrix} 0 & g_p \\ {}^t g_p & H_p \end{pmatrix}$$

of f enters the game. With the above notations and under the only permanent assumption $g_p \neq 0$, we have

$$\dim K_p = 4 - \operatorname{rank} \overline{H}_p. \tag{$*$}$$

In order to prove $(*)$, we consider the endomorphism

$$\overline{H}_p : \mathbb{R}^4 \longrightarrow \mathbb{R}^4, \quad \overline{v} \longmapsto \overline{H}_x \cdot \overline{v},$$

where $\overline{v} = \binom{v_0}{v}$ with $v_0 \in \mathbb{R}$ and $v \in \mathbb{R}^3$, and we define

$$\overline{K}_p := \operatorname{Ker} \overline{H}_p \in \mathbb{R}^4.$$

Now we show that the projection map

$$\xi : \overline{K}_p \longrightarrow K_p, \quad \overline{v} \longmapsto v,$$

defines an isomorphism. This implies $(*)$ and equivalently

$$\operatorname{rank} \overline{H}_p = \operatorname{rank} d_p \gamma + 2 ; \tag{$**$}$$

a result which remains true in a more general situation (2.4.2).

To simplify the notation we omit the subscripts x, p in g_x, H_x, and $\overline{H}_p$. Now

$$\overline{H} \cdot \overline{v} = \begin{pmatrix} g \cdot v \\ H \cdot v + v_0 g \end{pmatrix}.$$

Hence $\overline{v} \in \operatorname{Ker} \overline{H}$ implies $g \cdot v = 0$ and $H \cdot v = -v_0 g$, which means $v = \xi(\overline{v}) \in K$.

ξ is injective: If $v = 0$, then $H \cdot v = 0$ and $v_0 g = 0$, hence $v_0 = 0$ and $\overline{v} = 0$.

ξ is surjective: If $d_p\gamma(v) = 0$, then $g \cdot v = 0$ and $H \cdot v = \lambda g$ for some $\lambda \in \mathbb{R}$. Then

$$\overline{v} := \left(\begin{array}{c} -\lambda g \\ v \end{array} \right) \in \overline{K} \quad \text{and} \quad \xi(\overline{v}) = v \,.$$

Finally, relation ($**$) and the results of 0.2 imply the classical [Ei, Ch. I, §12]

Theorem. *Let $X = \{x \in \mathbb{R}^3 : f(x) = 0\}$ be a smooth surface defined by a differentiable function f and denote by $\overline{H}_x$ the extended Hesse matrix of f at $x \in X$.*

If X is developable around p, then $\det \overline{H}_x = 0$ around p.

If $\operatorname{rank} \overline{H}_x = 3$ around p, then X is developable around p.

0.4 Classification of Developable Surfaces in $\mathbb{R}^3$

In 0.1 we gave the three obvious examples of developable surfaces : cylinders, cones and tangent surfaces. Now we want to prove the classical result that locally there are no other examples.

We start with a local parameterization

$$\varphi : I \times J \longrightarrow U \subset X \subset \mathbb{R}^3 \,, \quad \varphi(s, t) = \alpha(s) + t\varrho(s) \,,$$

of a developable ruling. For the tangent plane we know

$$T_{\varphi(s,t)}X = \operatorname{span}\left\{\alpha'(s), \varrho(s)\right\}$$

and developability means $\varrho'(s) \in \operatorname{span}\left\{\alpha'(s), \varrho(s)\right\}$. To simplify the computation we use a ϱ with $\langle \varrho, \varrho \rangle = 1$, hence $\langle \varrho, \varrho' \rangle = 0$. Now we have to assume that

$$\text{either } \varrho'(s) = 0 \quad \text{or} \quad \varrho'(s) \neq 0 \quad \text{for every } s \in I \,. \tag{+}$$

In the first case, U is obviously part of a cylinder. In the second case, we extend φ to

$$\Phi : I \times \mathbb{R} \longrightarrow \mathbb{R}^3 \,, \quad \Phi(s, t) = \alpha(s) + t\varrho(s) \,,$$

which may have images outside of X for $t \notin I$; we show that there is a uniquely determined *critical directrix*, i.e. a map

$$\delta : I \longrightarrow \mathbb{R}^3 \quad \text{with} \quad \delta(s) = \alpha(s) + \mu(s)\varrho(s)$$

for some differentiable function μ on I, such that

$$\langle \delta', \varrho' \rangle = 0 \quad \text{on } I \,. \tag{*}$$

Of course, $\delta(I)$ may lie far outside U and even X.

Now $\delta' = \alpha' + \mu'\varrho + \mu\varrho'$; hence, (∗) implies

$$\langle \delta', \varrho' \rangle = \langle \alpha', \varrho' \rangle + \mu \langle \varrho', \varrho' \rangle = 0 \quad \text{and} \quad \mu := -\frac{\langle \alpha', \varrho' \rangle}{\langle \varrho', \varrho' \rangle}.$$

So μ is uniquely determined, and obviously δ, defined via this function μ, has the properties of a critical directrix. It is also easy to verify that δ does not depend on the choice of α and ϱ.

Now we have to take a closer look at the condition (∗). We know

$$\delta'(s) \in \text{span}\,\{\alpha'(s), \varrho(s)\} = \text{span}\,\{\alpha'(s), \varrho(s), \varrho'(s)\} = \text{span}\,\{\varrho(s), \varrho'(s)\};$$

hence, for every individual $s \in I$, we have

$$\text{either } \delta'(s) = 0 \quad \text{or} \quad \varrho(s) = \lambda(s)\delta'(s). \tag{++}$$

We have to make the assumption that one of these alternatives is valid for every $s \in I$. Then

- either $\Phi(I \times \mathbb{R}) \subset \mathbb{R}^3$ is a cone with vertex $\delta(I)$

- or $\Phi(I \times \mathbb{R}) \subset \mathbb{R}^3$ is the tangent surface of the smooth curve $\delta(I)$

In both cases, U is an open part of $\Phi(I \times \mathbb{R})$.

Since the additional assumptions (+) and (++) can be made on an open and dense set of X, we have proved the

Theorem. *If $X \subset \mathbb{R}^3$ is a smooth surface which has a developable ruling around every point, then for almost every point there is a neighborhood which is part of a cylinder, a cone, or a tangent surface.*

Of course the vertices of the cones or the critical curves of the tangent surfaces are never contained in X, since these points are singular.

So far we have made only local statements about developable surfaces. In $\mathbb{R}^3$, cylinders over smooth curves are the only obvious examples of smooth closed surfaces which are developable everywhere. Cones and tangent surfaces can be extended to closed surfaces, but they have singular points. One might try to puzzle together smooth parts of developable surfaces in a C^∞ way, in order to obtain a closed smooth surface. It took quite a long time to prove that this is not possible:

Cylinder Theorem. *The only closed smooth surfaces $X \subset \mathbb{R}^3$, such that the differential of the Gauß map has rank 1 everywhere, are cylinders.*

Proofs have been given in [Po], [HN] and [M]. HARTMAN and NIRENBERG [HN] proved the result more generally for hypersurfaces $X \subset \mathbb{R}^n$. In 2.2.9 and 2.3.5, we show by examples that the Cylinder Theorem is wrong if the Gauß map has rank greater than one.

0.5   Developable Surfaces in $\mathbb{P}_3(\mathbb{C})$

Developable differentiable surfaces can only be classified locally. In the algebraic case there is a classical global result, which is very easy to prove in the projective version [W].

We start with two different irreducible algebraic curves, $C_1, C_2 \subset \mathbb{P}_3 = \mathbb{P}_3(\mathbb{C})$, such that there is a Riemann surface S with holomorphic maps

$$\alpha : S \longrightarrow C_1 \quad \text{and} \quad \beta : S \longrightarrow C_2,$$

which are biholomorphic almost everywhere. Denote by

$$S' := \{s \in S : \alpha(s) \neq \beta(s)\}, \quad \text{and by} \quad \overline{\alpha(s)\beta(s)} \subset \mathbb{P}_3$$

the line joining these points for $s \in S'$. We use the fact that

$$X := \text{closure of} \bigcup_{s \in S'} \overline{\alpha(s)\beta(s)} \subset \mathbb{P}_3$$

is an irreducible algebraic surface, which is the union of a family of lines; we call such an X a *uniruled surface* (for details see Chapter 2). The curves $C_1, C_2 \subset X$ are called *directrices*, the lines $\overline{\alpha(s)\beta(s)}$ are called *generators*. As in the differentiable case, we call X *developable* if the tangent plane $\mathbb{T}_p X \subset \mathbb{P}_3$ to smooth points p of X is the same along every generator.

To get an analytic condition for developability, we use *local liftings* of α and β, i.e. an open $U \subset S$ and holomorphic maps

$$\widetilde{\alpha}, \widetilde{\beta} : U \longrightarrow \mathbb{C}^4 \quad \text{such that} \quad \alpha(s) = \mathbb{C}\widetilde{\alpha}(s) \quad \text{and} \quad \beta(s) = \mathbb{C}\widetilde{\beta}(s).$$

If we assume that X is smooth in $\alpha(s)$ and $\beta(s)$, then

$$\mathbb{T}_{\alpha(s)} X = \mathbb{P}(\text{span}\{\alpha(s), \beta(s), \alpha'(s)\}) \quad \text{and} \quad \mathbb{T}_{\beta(s)} X = \mathbb{P}(\text{span}\{\alpha(s), \beta(s), \beta'(s)\}),$$

where $'$ denotes the derivate with respect to s. Hence developability implies

$$\det(\widetilde{\alpha}, \widetilde{\beta}, \widetilde{\alpha}', \widetilde{\beta}') = 0 \quad \text{on } U. \tag{$*$}$$

Conversely, if this condition $(*)$ is satisfied, the tangent space must be the same along the generator. For details and the general situation see 2.2.4.

A uniruled surface $X \subset \mathbb{P}_3$ was constructed via two directrices

$$\alpha, \beta : S \longrightarrow X \subset \mathbb{P}_3,$$

but there are plenty of them. In general, a holomorphic map

$$\delta : S \longrightarrow X \subset \mathbb{P}_3$$

is called a *directrix* if $\delta(s) \in \overline{\alpha(s)\beta(s)}$ whenever $\alpha(s) \neq \beta(s)$. There are two cases of special interest:

1. There may exist a constant directrix map δ, i.e. a point $p \in X$ where every generator of X passes. In this case, X is a *cone* with *vertex* p. Of course, a plane X is a special case of a cone.

2. There may exist a directrix δ such that $\overline{\alpha(s)\beta(s)}$ is almost always the tangent line to $\delta(S)$ in $\delta(s)$. In this case, X is the *tangent surface* (or *tangent scroll*) of the curve $\delta(S)$.

Now we can formulate the classical result, which will be generalized in 2.2.8.

Theorem. *A uniruled surface $X \subset \mathbb{P}_3$ is developable iff X is either a cone or the tangent surface of some curve $C \subset X$. Moreover, the curve $C \subset X$ is uniquely determined.*

Proof. We use an open $U \subset S$ and liftings,

$$\widetilde{\alpha}, \widetilde{\beta}, \widetilde{\delta} : U \longrightarrow \mathbb{C}^4 \,,$$

of the directrices α, β, δ. First, there are holomorphic functions λ, μ on U such that

$$\widetilde{\delta} = \lambda\widetilde{\alpha} + \mu\widetilde{\beta} \,.$$

Now the conditions for δ to be "critical" are the following:

Case 1: $\widetilde{\delta} = \nu w$, where ν is holomorphic on U and $w \in \mathbb{C}^4$,

Case 2:   span $\{\widetilde{\delta}, \widetilde{\delta}'\}$ = span $\{\widetilde{\alpha}, \widetilde{\beta}\}$ almost everywhere on U.

In any case, there are holomorphic functions ϱ, σ on U such that

$$\widetilde{\delta}' = \varrho\widetilde{\alpha} + \sigma\widetilde{\beta} \,.$$

Since, on the other hand,

$$\widetilde{\delta}' = \lambda'\widetilde{\alpha} + \lambda\widetilde{\alpha}' + \mu'\widetilde{\beta} + \mu\widetilde{\beta}' \,,$$

we obtain

$$0 = (\lambda' - \varrho)\widetilde{\alpha} + \lambda\widetilde{\alpha}' + (\mu' - \sigma)\widetilde{\beta} + \mu\widetilde{\beta}' \,.$$

Since $(\lambda(s) : \mu(s)) \neq (0 : 0)$ for every $s \in U$, this implies the developability of cones and tangent surfaces via condition $(*)$ above.

The last equation also shows that a critical directrix δ is uniquely determined: we may assume that $\widetilde{\alpha}, \widetilde{\beta}, \widetilde{\alpha}'$ are linearly independent on an open subset of U since X is almost everywhere smooth. Hence, λ is uniquely determined for given μ, which implies that $(\lambda : \mu)$ is unique. Consequently, δ is locally and even globally unique.

This gives an immediate recipe for the proof of existence: if X is developable, we know

$$\det(\widetilde{\alpha}, \widetilde{\beta}, \widetilde{\alpha}', \widetilde{\beta}') = 0 \quad \text{on } U \,.$$

This implies that there are holomorphic functions λ, μ, λ^*, μ^* on U such that (1.1.3)

$$\lambda^*\widetilde{\alpha} + \lambda\widetilde{\alpha}' + \mu^*\widetilde{\beta} + \mu\widetilde{\beta}' = 0 \quad \text{on } U,$$

(+)

and we define

$$\widetilde{\delta} := \lambda\widetilde{\alpha} + \mu\widetilde{\beta}.$$

We compare the holomorphic functions

$$\widetilde{\delta} \wedge \widetilde{\delta}', \quad \widetilde{\alpha} \wedge \widetilde{\beta}: \quad U \longrightarrow \mathbb{C}^4 \wedge \mathbb{C}^4$$

and obtain, by using (+),

$$\begin{aligned}
\widetilde{\delta} \wedge \widetilde{\delta}' &= (\lambda\widetilde{\alpha} + \mu\widetilde{\beta}) \wedge (\lambda'\widetilde{\alpha} + \lambda\widetilde{\alpha}' + \mu'\widetilde{\beta} + \mu\widetilde{\beta}') \\
&= (\lambda(\mu' - \mu^*) - \mu(\lambda' - \lambda^*))\widetilde{\alpha} \wedge \widetilde{\beta} \\
&=: f(\widetilde{\alpha} \wedge \widetilde{\beta}).
\end{aligned}$$

Now we have to distinguish two cases for the holomorphic function f:

1. If $f = 0$, then $\widetilde{\delta} \wedge \widetilde{\delta}' = 0$, and $\widetilde{\delta}'(s) \in \mathbb{C}\widetilde{\delta}(s)$ for every $s \in S$. This implies (1.1.5) that

$$\widetilde{\delta} = vw$$

   for some vector $w \in \mathbb{C}^4$; hence, X is a cone with vertex $\mathbb{C}w \in \mathbb{P}_3$.

2. If $f \neq 0$, then span $\{\widetilde{\delta}, \widetilde{\delta}'\} = $ span $\{\widetilde{\alpha}, \widetilde{\beta}\}$ almost everywhere, and hence everywhere on U. This is the case of a tangent surface.

□

Example. Consider the rational curves

$$\alpha : \mathbb{P}_1 \longrightarrow \mathbb{P}_3, \quad (s_0 : s_1) \longmapsto (3s_0^2 : 2s_0s_1 : s_1^2 : 0),$$

$$\beta : \mathbb{P}_1 \longrightarrow \mathbb{P}_3, \quad (s_0 : s_1) \longmapsto (0 : s_0^2 : 2s_0s_1 : 3s_1^2),$$

and take the open set $U = \{(s_0 : s_1) \in \mathbb{P}_1 : s_0 \neq 0\}$ with coordinate $s = \frac{s_1}{s_0}$. Then

$$(\widetilde{\alpha}, \widetilde{\beta}, \widetilde{\alpha}', \widetilde{\beta}') = \begin{pmatrix} 3 & 0 & 0 & 0 \\ 2s & 1 & 2 & 0 \\ s^2 & 2s & 2s & 2 \\ 0 & 3s^2 & 0 & 6s \end{pmatrix},$$

and the columns of this matrix satisfy the relation

$$\widetilde{\alpha}' - 2\widetilde{\beta} + s\widetilde{\beta}' = 0.$$

Hence we can choose $\lambda = 1$, $\mu = s$ and obtain

$$\widetilde{\delta}(s) = \widetilde{\alpha} + s\widetilde{\beta} = (3, 3s, 3s^2, 3s^3).$$

It follows that the surface X generated by the directrices α and β is the tangent surface of the twisted cubic

$$\delta : \mathbb{P}_1 \longrightarrow \mathbb{P}_3, \quad (s_0 : s_1) \longmapsto (s_0^3 : s_0^2 s_1 : s_0 s_1^2 : s_1^3).$$

Chapter 1

Grassmannians

1.1 Preliminaries

1.1.1 Algebraic Varieties

The sole purpose of this section is to recall some well known concepts and facts, and to fix notations; our standard reference is [Sh].

Our topics are affine and projective algebraic varieties over the field $\mathbb{C}$ of complex numbers. An *algebraic set* $X \subset \mathbb{C}^N$ is the common set of zeroes of polynomials $f_1, \ldots, f_m \in \mathbb{C}[T_1, \ldots, T_N]$; for short we write

$$X = V(f_1, \ldots, f_m).$$

An algebraic set X is called *irreducible* if it is not the union of two algebraic sets X_1 and X_2 with $X_1 \not\subseteq X_2$ and $X_2 \not\subseteq X_1$. Every algebraic set X has a unique decomposition into *irreducible components*

$$X = X_1 \cup \ldots \cup X_r.$$

An irreducible algebraic set is called an *algebraic variety.*

An algebraic set $X \subset \mathbb{C}^N$ is called a *cone* if $p \in X$ and $\lambda \in \mathbb{C}$ implies $\lambda p \in X$. This is equivalent to the condition that X can be defined as the common set of zeroes of homogeneous polynomials.

If $X \subset \mathbb{C}^N$ and $Y \subset \mathbb{C}^M$ are algebraic sets, then a map

$$\varphi : X \longrightarrow Y$$

is called *regular* if there are polynomials $f_1, \ldots, f_M \in \mathbb{C}[T_1, \ldots, T_N]$ such that

$$\varphi(x) = (f_1(x), \ldots, f_M(x)) \quad \text{for every } x \in X.$$

There is a canonical map

$$\pi : \mathbb{C}^{N+1} \setminus \{0\} \longrightarrow \mathbb{P}_N(\mathbb{C}), \quad (x_0, \ldots, x_N) \longmapsto (x_0 : \ldots : x_N)$$

to the complex projective space $\mathbb{P}_N(\mathbb{C}) = \mathbb{P}(\mathbb{C}^{N+1})$. Since we only consider the field $\mathbb{C}$, we may write

$$\mathbb{P}_N = \mathbb{P}_N(\mathbb{C}).$$

A subset $X \subset \mathbb{P}_N$ is called (*projective*) *algebraic* if there is an algebraic cone $Y \subset \mathbb{C}^{N+1}$ with vertex $0 \in \mathbb{C}^{N+1}$, such that $X = \mathbb{P}(Y)$. Since Y is uniquely determined by X, we may write $Y = \widehat{X}$, and call $\widehat{X} \subset \mathbb{C}^{N+1}$ the *affine cone* of X; that is

$$X = \mathbb{P}(\widehat{X}) = \pi(\widehat{X} \backslash \{0\}) .$$

As in the case of an affine algebraic set; there is a decomposition into irreducible components in the projective case. These decompositions correspond in the following sense: if

$$\widehat{X} = \widehat{X}_1 \cup \ldots \cup \widehat{X}_r , \quad \text{then } X = X_1 \cup \ldots \cup X_r .$$

An irreducible algebraic set $X \subset \mathbb{P}_N$ is called a *projective algebraic variety*.

If X is algebraic in $\mathbb{C}^N$ or $\mathbb{P}_N$, then a subset $U \subset X$ is called *Zariski-open*, if $X \backslash U$ is algebraic. This determines the *Zariski-topology* of an algebraic set.

The definition of regular maps is more complicated in the projective case since the only polynomial functions on compact sets are constants. Assume $X \subset \mathbb{P}_N$ and $Y \subset \mathbb{P}_M$ are algebraic, then a map

$$\varphi : X \longrightarrow Y$$

is called *regular* if for every $p \in X$ there is a Zariski-open set $U \subset X$ with $p \in U$ and there are homogeneous polynomials $F_0, \ldots , F_M \in \mathbb{C}[T_0, \ldots , T_N]$ of the same degree such that

$$\varphi(x) = (F_0(x) : \ldots : F_M(x)) \quad \text{for every } x \in U .$$

In particular this condition includes the assumption

$$V(F_0, \ldots , F_M) \cap U = \emptyset .$$

In fact this condition is very restrictive; it is not difficult to prove that the only regular maps

$$\varphi : \mathbb{P}_N \longrightarrow \mathbb{P}_M$$

are constant maps in case $M < N$. In projective geometry rational maps are the more adequate concept (1.1.2). This reflects the geometric fact that projections have a center, where the map has no unique value. The points of the center are called points of indeterminacy of the rational map.

A consequence of elimination theory is the fundamental

Regular Mapping Theorem. *If $X \subset \mathbb{P}_N$ is an algebraic set and $\varphi : X \longrightarrow \mathbb{P}_M$ is regular, then $\varphi(X) \subset \mathbb{P}_M$ is an algebraic set.*

There are important relations between the dimensions of the algebraic sets involved in a regular map.

Dimension Theorem for Regular Maps. *If $X \subset \mathbb{P}_N$ is an algebraic variety, $\varphi : X \to \mathbb{P}_M$ is a regular map and $Y := \varphi(X)$, then the following holds:*

   a) Y is a variety and $\dim Y \leq \dim X$.

   b) If $y \in Y$ and $Z \subset \varphi^{-1}(y)$ is an irreducible component, then

$$\dim Z \geq \dim X - \dim Y .$$

   c) There is a non-empty Zariski-open subset $V \subset Y$ such that

$$\dim \varphi^{-1}(y) = \dim X - \dim Y \quad \text{for every } y \in V .$$

On several occasions we will need the following

Irreducibility Theorem. *Let $X \subset \mathbb{P}_N$ be an algebraic set and $\varphi : X \to \mathbb{P}_M$ a regular map. Assume:*

   a) $\varphi(X)$ is irreducible,

   b) for every $y \in \varphi(X)$ the fiber $\varphi^{-1}(y)$ is irreducible,

   c) all fibers $\varphi^{-1}(y)$ have the same dimension.

Then X is irreducible and $\dim X = \dim \varphi(X) + \dim \varphi^{-1}(y)$.

Many properties of varieties and maps are already determined by the following behavior of "first order": if X is an affine or projective algebraic set, then for every $p \in X$ there is the local ring $\mathcal{O}_{X,p}$ of germs of functions which are regular around p with the maximal ideal

$$\mathfrak{m}_{X,p} \subset \mathcal{O}_{X,p}$$

of functions vanishing at p. The vector space

$$T_p X := (\mathfrak{m}_{X,p} / \mathfrak{m}_{X,p}^2)^*$$

is called the *Zariski tangent space* of X at p.

If X is an affine or projective variety, then

   a) $\dim T_p X \geq \dim X$ for every $p \in X$,

   b) $\operatorname{Sing} X := \{p \in X : \dim T_p X > \dim X\} \subset X$ is an algebraic set.

Sing $X \subset X$ is called the *singular locus* . A point $p \in X$ is called *smooth* if $p \notin \text{Sing } X$, i.e. $\dim T_p X = \dim X$. The smooth part of X is denoted by $X_{\text{sm}} = X \setminus \text{Sing } X$.

If $\varphi : X \to Y$ is a regular map between affine or projective varieties, then for every $p \in X$ there is a canonical linear map

$$d_p \varphi : T_p X \longrightarrow T_{\varphi(p)} Y \,,$$

which is called the *differential of φ at p*. There is the important

Jacobian Criterion for Regular Maps. *If $\varphi : X \to Y$ is a regular surjective map between projective varieties, then there is a nonempty Zariski-open subset $U \subset X$ such that for every $p \in U$ we have:*

> *a)* $\dim X = \dim T_p X$,
>
> *b)* $\dim Y = \text{rank } d_p \varphi$,
>
> *c)* $\dim \varphi^{-1}(\varphi(p)) = \dim \text{Ker } d_p \varphi$.

For many geometrical considerations it is more adequate to consider the Zariski tangent space in an embedded form. For a variety $X \subset \mathbb{P}_N$ and $p \in X$ we consider the affine cone $\widehat{X}$ and some q over p with

$$0 \neq q \in \widehat{p} = \mathbb{C}q \subset \widehat{X} \subset \mathbb{C}^{N+1} \,.$$

Now the Zariski tangent space $T_q \widehat{X} \subset \mathbb{C}^{N+1}$ can be described as follows: if

$$\widehat{X} = V(F_1, \dots, F_m) \,,$$

where $F_1, \dots, F_m \in \mathbb{C}[T_0, \dots, T_N]$ are homogeneous, then

$$T_q \widehat{X} = \{v \in \mathbb{C}^{N+1} : \text{grad}_q F_k(v) = 0 \quad \text{for } k = 1, \dots, m\} \,.$$

We use EULER's formula: if $F \in \mathbb{C}[T_0, \dots, T_N]$ is homogenous of degree $d > 0$ then

$$\sum_{i=0}^{N} T_i \frac{\partial F}{\partial T_i} = d \cdot F \,.$$

As an immediate consequence we obtain $\mathbb{C}q \subset T_q \widehat{X}$. If we define

$$\mathbb{T}_p X := \mathbb{P}(T_q \widehat{X}) \subset \mathbb{P}_N \,,$$

this implies $p \in \mathbb{T}_p X$. We call $\mathbb{T}_p X \subset \mathbb{P}_N$ the *projective* (or *embedded*) *tangent space* of X at p. It is easy to see that

$$\dim \mathbb{T}_p X = \dim T_p X$$

for every $p \in X$.

1.1.2 Rational Maps

As we have seen in 1.1.1, the concept of regular maps is too restrictive in projective geometry, since points of indeterminacy are quite natural. This leads to the concept of rational maps, which are regular on a Zariski-open set and need not be single valued outside. More precisely, let $X \subset \mathbb{P}_N$ and $Y \subset \mathbb{P}_M$ be varieties (irreducibility is important in this context). Then a *rational map*

$$\varphi : X \dashrightarrow Y$$

is an equivalence class of regular maps

$$\varphi_U : U \longrightarrow Y$$

defined on some Zariski-open $U \subset X$, where φ_U and φ_V are equivalent if

$$\varphi_U = \varphi_V \quad \text{on } U \cap V .$$

In particular, there are homogeneous polynomials

$$f_0, \dots, f_M \in \mathbb{C}[T_0, \dots, T_N]$$

of the same degree such that $U := X \backslash V(f_0, \dots, f_M) \neq \emptyset$ and

$$\varphi(x) = (f_0(x) : \dots : f_M(x)) \in Y \quad \text{for every } x \in U .$$

Obviously, there is a maximal open set

$$\mathrm{Def}\, \varphi \subset X ,$$

where φ is regular; it is called the *domain of definition* of φ. The closed set

$$\mathrm{Cen}\, \varphi := X \backslash \mathrm{Def}\, \varphi$$

is called the *center* of φ. A rational map φ is regular iff $\mathrm{Def}\, \varphi = X$, i.e. $\mathrm{Cen}\,(\varphi) = \emptyset$. For a better understanding of the behavior of a rational map in its center, it is useful to consider its graph. For this purpose it is sufficient to consider the case $Y = \mathbb{P}_M$.

First, if $\varphi : X \to \mathbb{P}_M$ is a regular map, then its *graph*

$$\Gamma_\varphi := \{(x, y) \in X \times \mathbb{P}_M : y = \varphi(x)\} \subset X \times \mathbb{P}_M$$

is an algebraic set; it is the common set of zeroes of all bihomogeneous polynomials of the form

$$S_i f_j - S_j f_i \in \mathbb{C}[T_0, \dots, T_N ; S_0, \dots, S_M]$$

for all families of $(f_0, \dots, f_M)$ used in the local descriptions of φ. Since φ is a single valued map, the restriction of the canonical projection

$$\pi : \Gamma_\varphi \longrightarrow X$$

is biregular; in particular Γ_φ is a variety.

More generally, a variety $\Gamma \subset X \times \mathbb{P}_M$ is called a *rational graph* if the restriction of the canonical projection

$$\pi : \Gamma \longrightarrow X$$

is biregular almost everywhere, i.e. there is a Zariski-open subset $U \subset X$ such that the restriction of π

$$\pi^{-1}(U) \longrightarrow U$$

is biregular. Such a graph may be considered as the geometric picture of a rational map; it will be used to define images and inverse images of a rational map.

Proposition. *For every variety $X \subset \mathbb{P}_N$ there is a canonical one-to-one correspondence between rational maps*

$$\varphi : X \dashrightarrow \mathbb{P}_M$$

and rational graphs

$$\Gamma \subset X \times \mathbb{P}_M .$$

Proof. If we have a rational map φ, then there are polynomials $f_0, \ldots, f_M$ of the same degree defining a regular map φ_U on some open $U \subset X$. We consider the algebraic set

$$\Gamma^* := \{(x; y) \in X \times \mathbb{P}_M : y_i f_j(x) - y_j f_i(x) = 0 \text{ for all } i, j\} .$$

Over U this is the graph of φ_U, hence the projection

$$\Gamma_0 := \Gamma^* \cap \pi^{-1}(U) \longrightarrow U$$

is biregular. Now we define $\Gamma \subset \Gamma^*$ as the irreducible component of Γ^* containing Γ_0. It is well known that the equation

$$\Gamma = \text{closure of } \Gamma_0 \text{ in } \Gamma^* ,$$

is not only true in the Zariski-topology, but also in the topology of complex projective space [Wh$_1$, 3,2 I]. Obviously Γ is a rational graph.

Conversely, if we start with a rational graph $\Gamma \subset X \times \mathbb{P}_M \subset \mathbb{P}_N \times \mathbb{P}_M$, we consider its projections

$$\pi : \Gamma \longrightarrow \mathbb{P}_N \quad \text{and } \eta : \Gamma \longrightarrow \mathbb{P}_M .$$

By definition, there is a point $\gamma = (p, q) \in \Gamma$ such that the canonical map of local rings

$$\mathcal{O}_{X,p} \longrightarrow \mathcal{O}_{\Gamma,\gamma}$$

induced by π is an isomorphism. Now we consider the homogeneous coordinate rings

$$\mathbb{C}[T] = \mathbb{C}[T_0, \ldots, T_N] \quad \text{of} \quad \mathbb{P}_N , \quad \mathbb{C}[S] = \mathbb{C}[S_0, \ldots, S_M] \quad \text{of} \quad \mathbb{P}_M ,$$

and the canonical maps

$$\pi^* : \mathcal{O}_{\mathbb{P}_N, p} \longrightarrow \mathcal{O}_{\Gamma, \gamma} , \quad \eta^* : \mathcal{O}_{\mathbb{P}_M, q} \longrightarrow \mathcal{O}_{\Gamma, \gamma} .$$

Since π^* is surjective, we can find $f_{ij} \in \mathcal{O}_{\mathbb{P}_N, p}$ such that

$$\pi^* \left(f_{ij} \right) = \eta^* \left(\frac{S_i}{S_j} \right) \quad \text{if } S_j(\gamma) \neq 0 .$$

For fixed i, j the function f_{ij} is a quotient of polynomials of the same degree. These functions and the representations as quotients are far from being unique, but we can finally arrange

$$f_0, \dots , f_M \in \mathbb{C}[T]$$

of the same degree such that

$$\pi^* \left(\frac{f_i}{f_j} \right) = \eta^* \left(\frac{S_i}{S_j} \right) \quad \text{if } S_j(\gamma) \neq 0 ,$$

in particular $f_j(p) \neq 0$. These polynomials yield a regular map

$$\varphi_U : U := X \backslash V(f_0, \dots , f_M) \longrightarrow \mathbb{P}_M , \quad x \longmapsto (f_0(x) : \dots : f_M(x)) ,$$

having $\pi^{-1}(U) \subset \Gamma$ as its graph. $\square$

If $X \subset \mathbb{P}_N$ is a variety and

$$\varphi : X \dashrightarrow \mathbb{P}_M$$

is a rational map, then its graph $\Gamma_\varphi \subset X \times \mathbb{P}_M$ with canonical projections

$$\pi : \Gamma_\varphi \longrightarrow X \quad \text{and} \quad \eta : \Gamma_\varphi \longrightarrow \mathbb{P}_M$$

can be used to define images and inverse images: For a subvariety $X' \subset X$ we define the *image*

$$\varphi(X') := \eta(\pi^{-1}(X')) \subset \mathbb{P}_M ,$$

which is algebraic, but not necessarily a variety since $\pi^{-1}(X')$ need not be irreducible. For a variety $Y \subset \mathbb{P}_m$ the *inverse image* is

$$\varphi^{-1}(Y) := \pi(\eta^{-1}(Y)) \subset X ,$$

which is also algebraic.

The results stated in 1.1.1 for regular maps are true in modified form for rational maps since φ can be replaced by η if X is modified to Γ_φ.

Unlike regular maps, rational maps can be restricted or composed only under some precaution: if the map

$$\varphi : X \dashrightarrow \mathbb{P}_M$$

is rational and $X' \subset X$ is a subvariety, we assume $X' \cap \operatorname{Def} \varphi \neq \emptyset$. We consider

$$\Gamma_0' := \pi^{-1}(X' \cap \operatorname{Def} \varphi) \subset \Gamma_\varphi$$

and define

$$\Gamma_\varphi' := \text{closure of } \Gamma_0' \text{ in } \Gamma_\varphi .$$

Then $\varphi|X' : X' \dashrightarrow \mathbb{P}_M$ is the rational map determined by its graph Γ_φ'.

As an example, the reader may consider the rational map

$$\varphi : \mathbb{P}_2 \dashrightarrow \mathbb{P}_1 , \quad (x_0 : x_1 : x_2) \longmapsto (x_1 : x_2) ,$$

with center in $(1 : 0 : 0)$, and take as X' different kinds of curves passing the center. If X' is the cusp $x_1^3 = x_0 x_2^2$, then $\varphi|X'$ is rational and single valued in $(1 : 0 : 0)$ but not regular. If $X \subset \mathbb{P}_N$, $Y \subset \mathbb{P}_M$ are varieties and

$$\varphi : X \dashrightarrow \mathbb{P}_M , \quad \psi : Y \dashrightarrow \mathbb{P}_R$$

are rational maps, then there is a composite of rational maps

$$\psi \circ \varphi : X \dashrightarrow \mathbb{P}_R$$

if $\varphi(X) \subset Y$, but $\varphi(X) \not\subset \operatorname{Cen} \psi$. The construction is as above: ψ can be restricted to the subvariety $\varphi(X) \subset Y$.

1.1.3 Holomorphic Linear Combinations

In this paragraph, we collect some basic facts about holomorphic linear combinations and dependence.

Proposition. (CRAMER's Rule) *Assume we have a domain $S \subset \mathbb{C}$ and holomorphic maps*

$$\varrho, \varrho_1, \ldots , \varrho_n : S \longrightarrow \mathbb{C}^N$$

with the following properties:

1) There is a point $o \in S$ such that $\varrho_1(o), \ldots , \varrho_n(o)$ are linearly independent.

2) $\varrho(s), \varrho_1(s), \ldots , \varrho_n(s)$ are linearly dependent for every $s \in S$.

Then we can find holomorphic functions

$$\lambda, \lambda_1, \ldots , \lambda_n : S \longrightarrow \mathbb{C} \quad \text{such that } \lambda(o) \neq 0 \quad \text{and} \quad \lambda\varrho = \sum_{i=1}^n \lambda_i \varrho_i .$$

In particular, λ has only isolated zeroes on S.

In case $\varrho_1(s), \ldots , \varrho_n(s)$ are linearly independent for every $s \in S$, we can choose $\lambda = 1$.

Proof. We consider the holomorphic $(N \times n)$-matrix

$$A(s) := (\varrho_1(s), \ldots, \varrho_n(s))$$

with the given maps as columns and the inhomogeneous system of linear equations

$$A(s) \cdot \mu(s) = \varrho(s), \qquad (*)$$

where $\mu = {}^t(\mu_1, \ldots, \mu_n)$ and the μ_i are unknown functions; more precisely, the μ_i are elements of the quotient field of the integral domain of holomorphic functions on S.

According to assumption *1)*, we have a $(n \times n)$-submatrix $\overline{A}$ of A such that

$$\det \overline{A}(o) \neq 0.$$

If $\overline{\varrho}$ denotes the vector obtained from ϱ like $\overline{A}$ from A, we have a new system

$$\overline{A}(s) \cdot \mu(s) = \overline{\varrho}(s), \qquad (\overline{*})$$

and every solution μ of $(\overline{*})$ solves $(*)$. On the subset $\{s \in S : \det \overline{A}(s) \neq 0\} \subset S$ the solutions of $(\overline{*})$ can be computed by CRAMER's rule as

$$\mu_i := \frac{\det \overline{A}_i}{\det \overline{A}},$$

where $\overline{A}_i$ is the matrix $\overline{A}$ with the i-th column replaced by $\overline{\varrho}$. Consequently, if we define

$$\lambda := \det \overline{A} \quad \text{and} \quad \lambda_i := \det \overline{A}_i,$$

then these functions are holomorphic on the whole domain S and satisfy the required equation

$$\lambda \varrho = \sum_{i=1}^{n} \lambda_i \varrho_i.$$

In the case that $\varrho_1, \ldots, \varrho_n$ are everywhere linearly independent, the functions $\lambda_1, \ldots, \lambda_n$ on S such that $\varrho = \sum \lambda_i \varrho_i$ are uniquely determined. The above argument shows that they can be calculated as $\lambda_i = \mu_i$ and that they must be holomorphic. $\qquad \square$

1.1.4 Limit Direction of a Holomorphic Path

Given vector valued functions

$$\varrho_1, \ldots, \varrho_n : S \longrightarrow \mathbb{C}^N$$

on some parameter space S, the question is what can be said about the points $s \in S$ where the rank of $\varrho_1(s), \ldots, \varrho_n(s)$ is not maximal. The simplest case is a domain $S \subset \mathbb{C}$ and $n = 1$.

Extension Lemma. *Assume $S \subset \mathbb{C}$ is a domain containing the origin o,*

$$\varrho : S \longrightarrow \mathbb{C}^{N+1}$$

is holomorphic with $\varrho(s) \neq 0$ for $s \neq o$. Then the map

$$\mathbb{P}(\varrho) : S \setminus \{o\} \longrightarrow \mathbb{P}_N, \quad s \longmapsto \mathbb{C}\varrho(s),$$

has a unique holomorphic extension to S.

Geometrically, the path ϱ has a unique limit direction for $s = o$ even if $\varrho(o) = 0$.

Proof. If $\varrho = {}^t(\varrho_0, \dots, \varrho_N)$, we put $k_i := \mathrm{ord}_o\varrho_i$ for $i = 0, \dots, N$. Let k be the minimum of k_i, then we have for s near o

$$\varrho(s) = s^k \widetilde{\varrho}(s) \quad \text{with } \widetilde{\varrho}(o) \neq 0.$$

Now $\mathbb{P}(\widetilde{\varrho})$ is the desired extension, since $\mathbb{P}(\varrho)(s) = \mathbb{P}(\widetilde{\varrho})(s)$ for $s \neq o$. $\qquad \square$

The case of arbitrary n and more general S is postponed to paragraph 1.2.6.

1.1.5 Radial Paths

If $S \subset \mathbb{C}$ is open and $\varrho : S \to \mathbb{C}^N$ is holomorphic, then the path ϱ is called *radial* if there is a $v \in \mathbb{C}^N$ and a holomorphic function β on S such that

$$\varrho(s) = \beta(s)v \quad \text{for every } s \in S.$$

In this case and if $\beta(s) \neq 0$, then

$$\varrho'(s) = \beta'(s)v = \frac{\beta'(s)}{\beta(s)}\varrho(s).$$

If $\varrho \neq 0$, then this is equivalent to the fact that the induced function $\mathbb{P}(\varrho)$ is constant, i.e. the tangent ϱ' is always in the direction ϱ. This property is sufficient:

Lemma. *Assume $S \subset \mathbb{C}$ is a domain and $\varrho : S \to \mathbb{C}^N$ a holomorphic path such that*

$$\varrho'(s) = \alpha(s)\varrho(s) \quad \text{for every } s \in S,$$

where α is holomorphic on S. Then ϱ is radial.

Proof. If $\varrho(s) = 0$ for every s, then ϱ is trivially radial. Otherwise take $o \in S$ such that $v := \varrho(o) \neq 0$ and consider the system of complex differential equations

$$\varrho'(s) = \alpha(s)\varrho(s), \quad \varrho(o) = v. \tag{$*$}$$

The differential equation

$$\beta'(s) = \alpha(s)\beta(s), \quad \beta(o) = 1,$$

has a unique holomorphic solution β. Hence $\varrho = \beta v$ is the unique solution of $(*)$. $\qquad \square$

1.2   Plücker Coordinates

The material of this section is classical; standard references are [GH$_1$] and [H]. For the convenience of the reader, however, we recall the basic definitions and line out some technical details.

1.2.1 Local Coordinates

For $0 \leq n \leq N$, the set

$$\mathrm{Gr}(n, N) := \{V \subset \mathbb{C}^N : V \text{ is a vector subspace of dimension } n\}$$

is called the *Grassmannian* of $n-$*planes* in $\mathbb{C}^N$. For $n = 1$, this is just the projective space $\mathbb{P}_{N-1}$. From the projective point of view, it is adequate to consider

$$\mathbb{G}(n, N) := \{E \subset \mathbb{P}_N : E \text{ is a linear subspace of dimension } n\},$$

the set of n-planes in $\mathbb{P}_N$; we may identify

$$\mathbb{G}(n, N) = \mathrm{Gr}(n + 1, N + 1).$$

Plücker introduced coordinates for linear subspaces. In modern terminology, the set $\mathrm{Gr}(n, N)$ can be provided with the structure of a compact complex manifold embedded in a large projective space. The topological and analytic structure of $\mathrm{Gr}(n, N)$ reflects the concepts of continuous and holomorphic variation of linear subspaces. This is done as follows.

For a matrix $A \in \mathrm{M}\,(N \times n; \mathbb{C})$ we denote the columns by $a_1, \dots, a_n$. If

$$\mathrm{M}^*_{N,n} := \{A \in \mathrm{M}(N \times n; \mathbb{C}) : \mathrm{rank}\,A = n\},$$

then we have a surjective map

$$\Phi : \mathrm{M}^*_{N,n} \longrightarrow \mathrm{Gr}(n, N), \quad A \longmapsto \mathrm{span}\,\{a_1, \dots, a_n\},$$

such that

$$\Phi(A) = \Phi(B) \quad \text{iff } B = A \cdot S \quad \text{for some } S \in \mathrm{GL}(n; \mathbb{C}).$$

For a set $I = \{i_1, \dots, i_n\} \subset \{1, \dots, N\}$ in the order $1 \leq i_1 < \dots < i_n \leq N$ we denote by A_I the submatrix of A containing only the rows $i_1, \dots, i_n$, and consider the subset

$$U_I := \{V \in \mathrm{Gr}(n, N) : V \cap \mathrm{span}\,\{e_j : j \notin I\} = 0\}.$$

Obviously $\det A_I \neq 0$ iff $\Phi(A) \in U_I$. Consequently, for fixed I, the matrices generating U_I can be given in a standard form. The simplest case is $I = \{1, \dots, n\}$. For a matrix $B \in \mathrm{M}\,((N - n) \times n; \mathbb{C}) = \mathbb{C}^{n(N-n)}$, we denote

$$B^I := \begin{pmatrix} E_n \\ B \end{pmatrix} \quad \in \mathrm{M}^*_{N,n}.$$

For arbitrary I, the matrix B^I has the canonical basis $e_1, \ldots, e_n$ of $\mathbb{C}^n$ in the rows $i_1, \ldots, i_n$ and the rows of B in the remaining rows. By definition, the matrix B_I will always satisfy $\det\left(B^I\right)_I = 1$.

The resulting map

$$\Phi_I : \mathbb{C}^{n(N-n)} \longrightarrow U_I, \quad B \longmapsto \Phi(B^I),$$

is bijective for every I; therefore, we have a system of $\binom{N}{n}$ coordinates for $\mathrm{Gr}(n, N)$. A simple check shows that the resulting transition functions are biholomorphic. We conclude that the Grassmannians are complex manifolds and

$$\dim \mathrm{Gr}(n, N) = n(N - n).$$

1.2.2 The Plücker Embedding

Let $V \subset \mathbb{C}^N$ is a vector space of dimension n with bases $(v_1, \ldots, v_n)$ and $(w_1, \ldots, w_n)$. Let A and B be matrices with the base vectors as columns. Since $B = A \cdot S$ for some invertible S, we have

$$\det B_I = \det S \cdot \det A_I$$

for every $I = \{i_1, \ldots, i_n\}$. Since the n-minors of A and B are proportional, the map

$$\pi : \mathrm{Gr}(n, N) \longrightarrow \mathbb{P}\left(\textstyle\bigwedge^n \mathbb{C}^N\right), \quad \mathrm{span}\{v_1, \ldots, v_n\} \longmapsto \mathbb{C}(v_1 \wedge \ldots \wedge v_n),$$

is well defined. For $V \subset \mathbb{C}^N$ and $I = \{i_1, \ldots, i_n\}$ the minors

$$p_I(V) = p_{i_1 \ldots i_n}(V) := \det A_I$$

are called the *homogeneous Plücker coordinates of V*, and π is called *Plücker embedding*. This is justified by the following

Theorem. *The map*

$$\pi : \mathrm{Gr}(n, N) \longrightarrow \mathbb{P}\left(\textstyle\bigwedge^n \mathbb{C}^N\right)$$

defined above is holomorphic and injective.

Proof. We have to show that π is holomorphic on every coordinate neighborhood U_I. For simplicity we choose $I = \{1, \ldots, n\}$, then

$$B^I = \begin{pmatrix} 1 & & & & & 0 \\ & \ddots & & & & \\ & & 1 & & & \\ & & & \ddots & & \\ 0 & & & & & 1 \\ b_{n+1,1} & \cdots & & & \cdots & b_{n+1,n} \\ \vdots & & & & & \vdots \\ & & \cdots & b_{ij} & \cdots & \\ \vdots & & & & & \vdots \\ b_{N,1} & \cdots & & & \cdots & b_{N,n} \end{pmatrix}.$$

For every $J = \{j_1, \ldots, j_n\}$ the minor $\det\left(B^I\right)_J$ is a polynomial in the variables b_{ij} of $\mathbb{C}^{n(N-n)}$; hence π is holomorphic. Now fix a particular b_{ij} for $n + 1 \leq i \leq N$ and $1 \leq j \leq n$, then

$$\det\left(B^I\right)_J = \pm b_{ij} \quad \text{for } J := (I \backslash \{j\}) \cup \{i\} \,.$$

This proves the injectivity of π on U_I. For $V \notin U_I$ we know that $p_I(V) = 0$; hence, π is injective. $\square$

In 1.3.1 we shall give methods to reconstruct the vector space $V \subset \mathbb{C}^N$ from its $\binom{N}{n}$ homogeneous Plücker coordinates $p_I(V)$. Here we only note the following:

Let $(v_1, \ldots, v_n)$ be a basis of V and $\omega := v_1 \wedge \ldots \wedge v_n$, then for $v \in \mathbb{C}^N$ we have

$$v \in V \quad \Longleftrightarrow \quad v \wedge \omega = 0 \in \bigwedge^{n+1}\mathbb{C}^N \,.$$

1.2.3 Lines in $\mathbb{P}_3$

The most classical case of Plücker coordinates deals with lines in $\mathbb{P}_3$, i.e. we have $\mathbb{G}(1, 3) = \mathrm{Gr}(2, 4)$. Every 2-dimensional $V \subset \mathbb{C}^4$ has Plücker coordinates

$$(p_{12} : p_{13} : p_{14} : p_{23} : p_{24} : p_{34}) \in \mathbb{P}_5 \,.$$

Lemma. *Consider 2-dimensional spaces $V, W \subset \mathbb{C}^4$ with Plücker coordinates p_{ij}, q_{ij} and $X = \mathbb{P}(V), Y = \mathbb{P}(W) \subset \mathbb{P}_3$. Then $X \cap Y \neq \emptyset$ iff*

$$p_{12}q_{34} - p_{13}q_{24} + p_{14}q_{23} + p_{34}q_{12} - p_{24}q_{13} + p_{23}q_{14} = 0 \,. \tag{$*$}$$

Proof. If $V = \mathrm{span}\{v_1, v_2\}$, $W = \mathrm{span}\{w_1, w_2\}$, we consider the (4×4)-matrix

$$A := (v_1, v_2, w_1, w_2) \,.$$

Now

$$X \cap Y \neq \emptyset \Longleftrightarrow \dim(V \cap W) > 0 \Longleftrightarrow \dim(V + W) < 4 \Longleftrightarrow \det A = 0 \,.$$

The left side of the relation $(*)$ is just a generalized Laplace expansion of $\det A$. $\square$

The special case $X = Y$ yields the

Corollary. *The Plücker coordinates p_{ij} of any 2-dimensional $V \subset \mathbb{C}^4$ satisfy the relation*

$$p_{12}p_{34} - p_{13}p_{24} + p_{14}p_{23} = 0 \,.$$

This implies that the image $\pi(\mathbb{G}(1, 3)) \subset \mathbb{P}_5$ is contained in the *Plücker quadric*

$$Q := \{(p_{12} : p_{13} : p_{14} : p_{23} : p_{24} : p_{34}) \in \mathbb{P}_5 : p_{12}p_{34} - p_{13}p_{24} + p_{14}p_{23} = 0\} \,.$$

The gradient of this quadratic equation is

$$(p_{34}, -p_{24}, p_{23}, p_{14}, -p_{13}, p_{12}),$$

consequently, Q is smooth and irreducible. Since $\dim \mathbb{G}(1,3) = 4$, general arguments imply

$$\pi(\mathbb{G}(1,3)) = Q \subset \mathbb{P}_5.$$

1.2.4 The Plücker Image

We want to describe the image $\pi(\mathrm{Gr}(n,N)) \subset \mathbb{P}(\bigwedge^n \mathbb{C}^N)$ of the Plücker map π. If $\omega \in \bigwedge^n \mathbb{C}^N$, we consider the linear map

$$\wedge_\omega : \mathbb{C}^N \longrightarrow \bigwedge^{n+1} \mathbb{C}^N, \qquad v \longmapsto v \wedge \omega.$$

If $\omega \neq 0$ and $\mathbb{C}\omega$ is in the image under π, then we have seen in 1.2.2 that

$$\pi^{-1}(\mathbb{C}\omega) = \mathrm{Ker} \wedge_\omega;$$

in particular, $\dim \mathrm{Ker} \wedge_\omega = n$. Now we prove that this condition characterizes the image [H, 6.1].

Wedge Criterion. *For $0 \neq \omega \in \bigwedge^n \mathbb{C}^N$, the following conditions are equivalent:*

i) $\mathbb{C}\omega \in \pi(\mathrm{Gr}(n,N))$,

ii) *ω is totally decomposable, i.e. $\omega = v_1 \wedge \ldots \wedge v_n$ for some $v_1, \ldots, v_n \in \mathbb{C}^N$,*

iii) $\dim \mathrm{Ker} \wedge_\omega \geq n$.

Proof. Everything is obvious except for *iii) $\Rightarrow$ ii)*, and this is a consequence of the following elementary remark from multilinear algebra. This remark also shows that the case $\dim \mathrm{Ker} \wedge_\omega > n$ cannot occur if $\omega \neq 0$. $\square$

Remark. *Take a vector space V, a form $\omega \in \bigwedge^n V$, and linearly independent vectors $v_1, \ldots, v_k \in V$ such that*

$$v_i \wedge \omega = 0 \in \bigwedge^{n+1} V \quad for\ i = 1, \ldots, k,$$

where $0 \leq k \leq n \leq \dim V$. Then there is a form $\xi \in \bigwedge^{n-k} V$ such that

$$\omega = v_1 \wedge \ldots \wedge v_k \wedge \xi.$$

Proof of the Remark. Take a basis $(v_1, \ldots, v_k, v_{k+1}, \ldots, v_N)$ of V, consider the corresponding basis (v_I) with $v_I = v_{i_1} \wedge \ldots \wedge v_{i_n}$ of $\bigwedge^n V$, and represent

$$\omega = \sum_I \omega_I v_I.$$

Now $v_i \wedge \omega = 0$ implies $\omega_I = 0$ if $i \notin I$; thus, $\{1, \dots, k\} \subset I$ whenever $\omega_I \neq 0$. $\square$

The advantage of this criterion is that condition *iii)* can be expressed in different terms. For example, we may consider the linear map

$$\wedge : \textstyle\bigwedge^n \mathbb{C}^N \longrightarrow \text{Hom}\,(\mathbb{C}^N, \textstyle\bigwedge^{n+1}\mathbb{C}^N), \qquad \omega \longmapsto \wedge_\omega.$$

Since $\dim \text{Ker} \wedge_\omega \geq n$ iff $\text{rank} \wedge_\omega \leq N - n$, we use the fact that

$$X = \{\varphi \in \text{Hom}(\mathbb{C}^N, \textstyle\bigwedge^{n+1}\mathbb{C}^N) : \text{rank}\,\varphi \leq N - n\}$$

is an algebraic subset, described by the vanishing of all $(N - n + 1)$-minors of the $\left(\binom{N}{n+1} \times N\right)$-matrix representing φ. Since the map $\wedge$ is linear, we obtain the

Corollary. *The Plücker image $\pi(\text{Gr}(n, N)) \subset \mathbb{P}(\bigwedge^n \mathbb{C}^N)$ is an algebraic set and can be described by polynomials of degree $N - n + 1$ in the Plücker coordinates.*

As an example, let us compute these (cubic) polynomials in case $N = 4$ and $n = 2$. They cannot generate the ideal, which contains the quadratic polynomial of 1.2.3. If

$$\omega = \sum_{1 \leq i < j \leq 4} p_{ij} e_i \wedge e_j$$

and the basis of $\bigwedge^3 \mathbb{C}^4$ is in lexicographic order, then the matrix A_ω representing $\wedge_\omega$ is

$$\begin{pmatrix} p_{23} & -p_{13} & p_{12} & 0 \\ p_{24} & -p_{14} & 0 & p_{12} \\ p_{34} & 0 & -p_{14} & p_{13} \\ 0 & p_{34} & -p_{24} & p_{23} \end{pmatrix},$$

and the sought-after polynomials are the 3-minors of this matrix.

In general, if $\omega = (p_I)$, the matrix A_ω representing $\wedge_\omega$ has entries a_{Ji}, where

$$J = \{j_1, \dots, j_{n+1}\} \text{ with } 1 \leq j_1 < \dots < j_{n+1} \leq N \text{ and } i \in \{1, \dots, N\}.$$

It is easy to see that

$$a_{Ji} = \begin{cases} 0 & \text{if } i \notin J \\ \sigma p_I & \text{if } i \in J \quad \text{and } J = I \cup \{i\}, \end{cases}$$

where $\sigma = \text{sgn}(i, i_1, \dots, i_n)$ if $I = \{i_1, \dots, i_n\}$.

1.2.5 Plücker Relations

By using some more multilinear algebra we compute the classical quadratic equations for the Plücker image [H, 6.1].

To simplify the notations for a while, we write $V = \mathbb{C}^N$. We have the dual pairings

$$\textstyle\bigwedge^n V \times \bigwedge^{N-n} V \quad \longrightarrow \quad \bigwedge^N V \cong \mathbb{C}, \quad (\omega, \xi) \longmapsto \omega \wedge \xi \,,$$

$$\textstyle\bigwedge^{N-n} V^* \times \bigwedge^{N-n} V \longrightarrow \mathbb{C}, \quad (\varphi_1 \wedge \ldots \wedge \varphi_{N-n},\ v_1 \wedge \ldots \wedge v_{N-n}) \longmapsto \det(\varphi_i(v_j)) \,.$$

Unfortunately, the isomorphism $\bigwedge^N V \cong \mathbb{C}$ is not canonical. Using the canonical basis $(e_1, \ldots, e_N)$ of V, we fix one isomorphism by

$$\textstyle\bigwedge^N V \longrightarrow \mathbb{C}, \quad \lambda(e_1 \wedge \ldots \wedge e_N) \longmapsto \lambda \,.$$

Then the dual pairings induce the following isomorphisms:

$$\textstyle\bigwedge^n V \xrightarrow{\ \cong\ } (\bigwedge^{N-n} V)^* \xrightarrow{\ \cong\ } \bigwedge^{N-n} V^* \,.$$

If $\omega \in \bigwedge^n V$, then the corresponding element in $\bigwedge^{N-n} V^*$ will be denoted by ω^*.

In order to write this isomorphism out in coordinates, let $(e_1^*, \ldots, e_N^*)$ be the dual of the canonical basis, the corresponding basis for

$$\textstyle\bigwedge^n V \quad \text{is } (e_I)_I \quad \text{and for} \quad \bigwedge^{N-n} V^* \quad \text{is } (e_J^*)_J \,,$$

where $I = \{i_1, \ldots, i_n\} \subset \{1, \ldots, N\}$ and $J = \{j_1, \ldots, j_{N-n}\} \subset \{1, \ldots, N\}$ are strictly ordered index sets. Further we define

$$I^* := \{1, \ldots, N\} \setminus I = \{k_1, \ldots, k_{N-n}\} \quad \text{and } \sigma(I) = \operatorname{sgn}(k_1, \ldots, k_{N-n}, i_1, \ldots, i_n) \,,$$

then the above isomorphism is determined by

$$(e_I)^* = \sigma(I) \cdot e_{I^*}^* \,.$$

In particular, if $(v_1, \ldots, v_N)$ is any basis of $\mathbb{C}^N$, then

$$(v_1 \wedge \ldots \wedge v_n)^* = \alpha(v_{n+1}^* \wedge \ldots \wedge v_N^*)$$

for some $\alpha \in \mathbb{C}^*$.

For $\omega \in \bigwedge^n V$ we have the two homomorphisms

$$\wedge_\omega : V \quad \longrightarrow \quad \textstyle\bigwedge^{n+1} V \,, \qquad v \longmapsto v \wedge \omega \,,$$

$$\wedge_{\omega^*} : V^* \quad \longrightarrow \quad \textstyle\bigwedge^{N-n+1} V^* \,, \qquad \varphi \longmapsto \varphi \wedge \omega^* \,.$$

With respect to the canonical bases (in lexicographical order for the exterior products) $\wedge_\omega$ and $\wedge_{\omega^*}$ are given by matrices

$$A_\omega \in M\left(\binom{N}{n+1} \times N; \mathbb{C}\right) \quad \text{and} \quad B_{\omega^*} \in M\left(\binom{N}{N-n+1} \times N; \mathbb{C}\right).$$

Recall that for a vector subspace $W \subset V$ the *annihilator* is denoted by $W^\circ \subset V^*$. Now we can formulate the

Main Lemma. *For* $\omega = \sum_I p_I e_I \in \bigwedge^n \mathbb{C}^N$, *the following conditions are equivalent:*

  i) ω *is totally decomposable,*

  ii) ω^* *is totally decomposable,*

  iii) $\operatorname{Ker} \wedge_{\omega^*} = (\operatorname{Ker} \wedge_\omega)^\circ$,

  iv) $A_\omega \cdot {}^t B_{\omega^*} = 0$,

  v) for every pair $J = \{j_1, \dots, j_{n+1}\}$, $K = \{k_1, \dots, k_{n-1}\} \subset \{1, \dots, N\}$ *of strictly ordered index sets*

$$\sum_{l=1}^{n+1} (-1)^l p_{j_1 \dots \widehat{j_l} \dots j_{n+1}} \, p_{k_1 \dots j_l \dots k_{n-1}} = 0.$$

As usual, ˆ over an index means that it is omitted. The quadratic equations in *v)* are called *Plücker relations.* Together with the Wedge Criterion of 1.2.4 we obtain the

Corollary. *The Plücker image* $\pi(\operatorname{Gr}(n, N)) \subset \mathbb{P}(\bigwedge^n \mathbb{C}^N)$ *is the intersection of quadrics (i.e. of quadratic hypersurfaces).*

Proof of the Main Lemma. Along with the equivalence of *i)* and *ii)*, which has been shown above, the Wedge Criterion implies

$$\dim \operatorname{Ker} \wedge_\omega = n \quad \text{and} \quad \dim \operatorname{Ker} \wedge_{\omega^*} = N - n;$$

hence, it suffices to show $\operatorname{Ker} \wedge_{\omega^*} \subset (\operatorname{Ker} \wedge_\omega)^\circ$. Take $\varphi \in \operatorname{Ker} \wedge_{\omega^*}$ and $v \in \operatorname{Ker} \wedge_\omega$; we have to show $\varphi(v) = 0$. There is a basis $(v_1, \dots, v_n)$ of $\mathbb{C}^N$ such that

$$\omega = v_1 \wedge \dots \wedge v_n \quad \text{and} \quad \omega^* = v_{n+1}^* \wedge \dots \wedge v_N^*.$$

Since $\varphi \in \operatorname{span}\{v_{n+1}^*, \dots, v_N^*\}$ and $v \in \operatorname{span}\{v_1, \dots, v_n\}$, we get $\varphi(v) = 0$.

For arbitrary ω, we have $\dim \operatorname{Ker} \wedge_\omega \leq n$ and $\dim \operatorname{Ker} \wedge_{\omega^*} \leq N - n$; therefore, condition *iii)* implies that there is equality in both cases. This yields *i)* and *ii)*.

By duality, condition *iii)* is equivalent to

$$\operatorname{Im} \wedge_\omega^* = (\operatorname{Im} \wedge_{\omega^*}^*)^\circ;$$

expressed in matrices, this is iv). Condition iv) means that the rows of A_ω and B_ω^* are mutually orthogonal. Except for the signs, the entries of A_ω and B_{ω^*} are easy to compute. By 1.2.4 and the definition of ω^* we have

$$a_{Ji} = \begin{cases} 0 & \text{if } i \notin J \\ \pm p_{J\setminus i} & \text{if } i \in J \end{cases} \qquad b_{Ki} = \begin{cases} 0 & \text{if } i \in K \\ \pm p_{K\cup i} & \text{if } i \notin K. \end{cases}$$

By chasing the signs one can see that

$$\sum_{i=1}^{N} a_{Ji} b_{Ki} = 0$$

is just the expression in v).   □

Our standard example is $n = 2$, $N = 4$. A_ω has been computed in 1.2.4, and

$$B_{\omega^*} = \begin{pmatrix} p_{14} & p_{24} & p_{34} & 0 \\ -p_{13} & -p_{23} & 0 & p_{34} \\ p_{12} & 0 & -p_{23} & -p_{24} \\ 0 & p_{12} & p_{13} & p_{14} \end{pmatrix}.$$

The product of the first rows of A_ω and B_{ω^*} corresponding to $J = \{1, 2, 3\}$ and $K = \{4\}$ gives

$$p_{12}p_{34} - p_{13}p_{24} + p_{14}p_{23} = 0.$$

Other combinations give either the same relation or trivial ones.

1.2.6 Systems of Vector Valued Functions

In 1.1.4 we proved a simple extension lemma for holomorphic curves, which uses projective space. By means of Grassmannians and rational maps, we can now handle the general case.

Extension Lemma. *Assume $Y \subset \mathbb{C}^m$ is an affine variety and*

$$\varrho_1, \ldots, \varrho_n : Y \longrightarrow \mathbb{C}^N$$

are regular maps. We define

$$r(y) := \text{rank}\,(\varrho_1(y), \ldots, \varrho_n(y)), \quad r := \max\{r(y) : y \in Y\}, \quad \text{and}$$

$$Y^* := \{y \in Y : r(y) = r\}.$$

Then there is a uniquely determined rational map

$$\varrho : Y \dashrightarrow \text{Gr}(r, N)$$

such that $\varrho(y) = \text{span}\,\{\varrho_1(y), \ldots, \varrho_n(y)\}$ for $y \in Y^*$.

In case Y is one-dimensional and smooth, ϱ is regular.

Proof. Take a point $q \in Y$ such that $r(q) = r$, and assume

$$(\varrho_1(q), \ldots, \varrho_r(q)) \quad \text{is a basis of} \quad \text{span}\,\{\varrho_1(q), \ldots, \varrho_n(q)\}\,.$$

We consider the $(N \times r)$-matrix

$$A(y) = (\varrho_1(y), \ldots, \varrho_r(y))$$

with the indicated vectors as columns and $Y^\star := \{y \in Y : \text{rank}\,A(y) = r\}$. For every strictly ordered index set $I = \{i_1, \ldots, i_r\} \subset \{1, \ldots, N\}$, we have a minor $\det A_I$ of A; this induces a regular map

$$\tilde{\varrho} : Y \longrightarrow \textstyle\bigwedge^r \mathbb{C}^N\,, \quad y \longmapsto \varrho_1(y) \wedge \ldots \wedge \varrho_r(y) = (\det A_I(y))_I\,.$$

For $y \in Y^\star$, we have

$$\mathbb{C}\tilde{\varrho}(y) = \text{span}\,\{\varrho_1(y), \ldots, \varrho_r(y)\} \in \text{Gr}(r, N) \subset \mathbb{P}(\textstyle\bigwedge^r \mathbb{C}^N)\,.$$

In order to construct the graph Γ_ϱ of ϱ, we use homogeneous coordinates z_I in $\mathbb{P}(\bigwedge^r \mathbb{C}^N)$, and we define the algebraic set

$$\Gamma := \{(y; z) \in Y \times \mathbb{P}(\textstyle\bigwedge^r \mathbb{C}^N) : z_I \det A_J(y) = z_J \det A_I(y) \quad \text{for all pairs } I, J\}\,.$$

Then $\Gamma_\varrho \subset Y \times \text{Gr}(r, N)$ is defined as the component of Γ containing the graph of the regular map

$$\varrho^\star : Y^\star \longrightarrow \text{Gr}(r, N)\,, \quad y \longmapsto \mathbb{C}\tilde{\varrho}(y)\,.$$

Obviously, Γ depends on the choice of the system $\varrho_1, \ldots, \varrho_r$, with maximal rank somewhere within $\varrho_1, \ldots, \varrho_n$; but the component $\Gamma_\varrho \subset \Gamma$ is independent of this choice. Hence ϱ is regular on Y^*, not only on $Y^\star$. For $y \in Y \backslash Y^*$, the vector space

$$\text{span}\,\{\varrho_1(y), \ldots, \varrho_n(y)\}$$

has dimension less than r; hence, it lies in a different Grassmannian. The image

$$\varrho(y) \subset \text{Gr}(r, N)$$

is an algebraic subset consisting of all limit points of nearby images. The simplest case for this phenomenon is a zero of a single vector valued function.

In case $Y \subset \mathbb{C}$ is a domain, every rational map to a projective space is regular (1.1.4). $\square$

1.3   Incidences and Duality

In this section, we provide some technical tools concerning the relations between different Grassmannians.

1.3.1 Equations and Generators in Terms of Plücker Coordinates

For later use, we show how linear equations or generators for a vector space can be obtained in terms of its Plücker coordinates. If $V = \mathbb{C}^N$, then we consider

$$\mathrm{Gr}(n, N) \subset \mathbb{P}(\textstyle\bigwedge^n V)$$

as a subset via the Plücker embedding 1.2.2. For a subspace $W \subset V$ with base $(w_1, \dots, w_n)$ we put

$$\omega := w_1 \wedge \dots \wedge w_n = (p_I(W))_I \subset \textstyle\bigwedge^n V.$$

In paragraph 1.2.4, we considered the linear map

$$\wedge_\omega : V \longrightarrow \textstyle\bigwedge^{n+1} V, \quad v \longmapsto v \wedge \omega,$$

and we saw that $W = \mathrm{Ker}\, \wedge_\omega$. This yields a system of linear equations as follows. For $v \in V$ we consider the matrix

$$A^v := (v, w_1, \dots, w_n) \in M(N \times (n+1), \mathbb{C})$$

with the indicated vectors as columns. Now

$$v \in \mathrm{Ker}\, \wedge_\omega \iff \mathrm{rank}\, A^v < n + 1.$$

This last condition can be expressed by the vanishing of all $(n + 1)$-minors of A^v. If $v = {}^t(x_1, \dots, x_N)$, we can expand every minor by the first column, yielding the following system of $\binom{N}{n+1}$ linear equations for $W \subset V = \mathbb{C}^N$:

$$\sum_{j=1}^{n+1} (-1)^{j+1} p_{i_1 \dots \widehat{i_j} \dots i_{n+1}}(W) x_{i_j} = 0$$

for every set of indices $1 \le i_1 < \dots < i_j < \dots < i_{n+1} \le N$.

The reader may write down these equations explicitly in simple cases, e.g. $n = 2$ and $N = 4$. Alternatively, one might directly use the matrix A_ω defined in 1.2.5 to obtain the above system.

In order to obtain generators of W, we use the linear map

$$\wedge_{\omega^*}^* : \textstyle\bigwedge^{N-n+1} V \longrightarrow V$$

dual to the map $\wedge_{\omega^*}$ defined in 1.2.5. The Main Lemma implies

$$\mathrm{Ker}\, \wedge_\omega = (\mathrm{Ker}\, \wedge_{\omega^*})^\circ = \mathrm{Im}\, \wedge_{\omega^*}^*;$$

hence, $W = \mathrm{Im}\, \wedge_{\omega^*}^*$. So W is generated by the $\binom{N}{n-1}$ rows of B_{ω^*}, and we know that the entries of B_{ω^*} are Plücker coordinates up to a sign.

1.3.2 Flag Varieties

In this paragraph, we fix the notation $V = \mathbb{C}^N$ and consider two vector subspaces $U, W \subset V$, not necessarily of the same dimension. The question is, how different kinds of relations between U and W, for instance

$$U \subset W, \quad U \cap W \neq \{0\}, \quad \dim(U \cap W) \geq k$$

can be expressed in terms of the Plücker coordinates of U and W. In the above cases this is possible in polynomial form, which gives rise to new algebraic varieties.

If $0 \leq m \leq n \leq N$ we define the *flag variety*

$$\text{Flag}\,(m, n; N) := \{(U, W) \in \text{Gr}(m, N) \times \text{Gr}(n, N) : U \subset W\}.$$

In the projective notation, we write

$$\mathbb{F}(m, n; N) := \text{Flag}\,(m + 1, n + 1; N + 1).$$

Proposition. *The flag variety*

$$\text{Flag}\,(m, n; N) \subset \mathbb{P}(\textstyle\bigwedge^m V) \times \mathbb{P}(\textstyle\bigwedge^n V)$$

is an algebraic variety.

Proof. Take

$$\eta = (p_I)_I \in \textstyle\bigwedge^m V \quad \text{and} \quad \omega = (q_J)_J \in \textstyle\bigwedge^n V,$$

such that $U = \mathbb{C}\eta$ and $W = \mathbb{C}\omega$. Our goal is to express the relation $U \subset W$ in terms of the Plücker coordinates p_I and q_J of U and W.

According to 1.3.1, we have a system $u_1, \ldots, u_k$ of generators of U with $k = \binom{N}{m-1}$, where the components of the u_i are Plücker coordinates up to a sign. Now

$$U \subset W \iff u_i \wedge \omega = 0 \in \textstyle\bigwedge^{n+1} V \quad \text{for } i = 1, \ldots, k.$$

This yields $\binom{N}{m-1}\binom{N}{n+1}$ bilinear equations in the p_I and q_J generalizing the Plücker relations from 1.2.5; hence, Flag $(m, n; N)$ is algebraic.

The projection to a Grassmannian has linear fibers of constant dimension, this implies the irreducibility of the flag variety (1.1.1). $\qquad\square$

Corollary. *For $0 \leq k \leq m, n \leq N$, the set*

$$I(k; m, n; N) = \{(U, W) \in \text{Gr}(m, N) \times \text{Gr}(n, N) : \dim(U \cap W) \geq k\}$$

is an algebraic variety.

Proof. Consider the variety

$$\{(U', U, W', W) \in \text{Gr}(k, N) \times \text{Gr}(m, N) \times \text{Gr}(k, N) \times \text{Gr}(n, N) :$$

$$U' \subset U, W' \subset W, U' = W'\} \subset \text{Flag}\,(k, m; N) \times \text{Flag}\,(k, n; N)$$

and project it to $\text{Gr}(m, N) \times \text{Gr}(n, N)$. $\qquad\square$

1.3.3 Duality of Grassmannians

The elementary concept of duality from linear algebra and geometry has a natural extension to Grassmannians. As in the elementary case, the geometrical interpretation of this formal procedure may sometimes be quite surprising.

We start with a finite dimensional complex vector space V. For $0 \leq n \leq \dim V$, we have the Grassmannian

$$\mathrm{Gr}(n, V) = \{U \subset V : \dim U = n\}.$$

If V^* denotes the dual vector space of linear forms on V, then subspaces $U \subset V$ and $W \subset V^*$ have *annihilators*

$$U^\circ = \{\varphi \in V^* : \varphi(U) = 0\} \subset V^* \quad \text{and}$$

$$W^\circ = \{v \in V : W(v) = 0\} \subset V$$

of complementary dimensions. Moreover,

$$(U^\circ)^\circ = U \quad \text{and } (W^\circ)^\circ = W.$$

Expressed in terms of Grassmannians, this yields the

Duality Theorem. *For $0 \leq n \leq N = \dim V$, the map*

$$\mathcal{D} : \mathrm{Gr}(n, V) \longrightarrow \mathrm{Gr}(N - n, V^*), \quad U \longmapsto U^\circ,$$

is biregular with inverse

$$\mathcal{D}^* : \mathrm{Gr}(N - n, V^*) \longrightarrow \mathrm{Gr}(n, V), \quad W \longmapsto W^\circ.$$

Proof. We use the almost canonical isomorphism

$$\bigwedge^n V \longrightarrow \bigwedge^{N-n} V^*, \quad \omega \mapsto \omega^*,$$

from 1.2.5, which yields a canonical biregular map

$$\mathbb{P}(\bigwedge^n V) \longrightarrow \mathbb{P}(\bigwedge^{N-n} V^*).$$

Now, obviously, $\mathcal{D}$ is just its restriction to the Grassmannians. □

1.3.4 Dual Projective Spaces

From an abstract point of view, the duality isomorphism $\mathcal{D}$ is absolutely innocent, but for practical computations it creates sign problems. Let us consider $V = \mathbb{C}^{N+1}$, and

$$\mathbb{G}(n, N) := \mathrm{Gr}(n + 1, \mathbb{C}^{N+1}), \quad \mathbb{G}^*(m, N) := \mathrm{Gr}(m + 1, (\mathbb{C}^{N+1})^*),$$

since we want to return to projective subspaces. In this context the duality isomorphisms are denoted by

$$\mathcal{D}_{n,N} : \mathbb{G}(n, N) \longrightarrow \mathbb{G}^*(N - n - 1, N), \quad \mathcal{D}^*_{m,N} : \mathbb{G}^*(m, N) \longrightarrow \mathbb{G}(N - m - 1, N).$$

By the above theorem, $\mathcal{D}^*_{N-n-1,N}$ is the inverse of $\mathcal{D}_{n,N}$.

The simplest nontrivial case is $n = N - 1$. Then we have two kinds of *dual projective spaces*:

$$\mathbb{P}^\vee_N := \mathbb{G}(N - 1, N) \quad \text{and} \quad \mathbb{P}^*_N := \mathbb{G}^*(0, N) = \mathbb{P}((\mathbb{C}^{N+1})^*),$$

with an isomorphism

$$\mathcal{D}_{N-1,N} : \mathbb{P}^\vee_N \longrightarrow \mathbb{P}^*_N, \quad H \longmapsto \mathbb{C}\varphi,$$

associating its linear equations in $(\mathbb{C}^{N+1})^*$ to every hyperplane $H \subset \mathbb{P}_N$. We want to compare the Plücker coordinates $(p_0 : \ldots : p_N)$ of H with the homogeneous coordinates $(a_0 : \ldots : a_N)$ of $\mathbb{C}\varphi$. Take $U \subset \mathbb{C}^{N+1}$ with $H = \mathbb{P}(U)$, let

$$A := (v_1, \ldots, v_N)$$

be the $((N + 1) \times N)$-matrix with a basis of U as columns, and denote by A_i the submatrix with row i missing. Then by paragraph 1.2.2

$$p_i = \det A_i \quad \text{and} \quad a_i = (-1)^i \det A_i \quad \text{for } i = 0, \ldots, N;$$

where $p_i := p_{0\ldots\hat{i}\ldots N}(U)$.

For general n the relation between the Plücker coordinates in $\mathbb{G}(n, N)$ and $\mathbb{G}^*(N-n-1, N)$ is determined by the expression of the isomorphism

$$\bigwedge^{n+1} \mathbb{C}^{N+1} \longrightarrow \bigwedge^{N-n}(\mathbb{C}^{N+1})^*$$

in terms of the canonical bases (1.2.5).

1.4 Tangents to Grassmannians

1.4.1 Tangents to Projective Space

For later use we need a good description of the tangent space to a Grassmannian in an arbitrary point. The simplest case is projective space, where the tangent bundle can be described via the EULER *sequence* [GH₁, p. 409]. We first take a look at this case. Denote by

$$\pi : \mathbb{C}^{N+1} \setminus 0 \to \mathbb{P}_N, \quad y \mapsto \mathbb{C}y = L,$$

the canonical projection with its differential

$$d_y\pi : T_y(\mathbb{C}^{N+1}) \longrightarrow T_L\mathbb{P}_N.$$

The kernel of $d_y\pi$ is just the line L of radial tangent vectors, this gives the exact sequence

$$0 \longrightarrow L \longrightarrow T_y(\mathbb{C}^{N+1}) \longrightarrow T_L\mathbb{P}_N \longrightarrow 0 \, .$$

Now, the canonical isomorphism $T_y(\mathbb{C}^{N+1}) = \mathbb{C}^{N+1}$ can be viewed as the isomorphism

$$\widehat{\chi} : T_y\mathbb{C}^{N+1} \longrightarrow \mathrm{Hom}\,(L, \mathbb{C}^{N+1})\,, \quad \tau \longmapsto \widehat{\chi}_\tau\,, \quad \text{defined by} \quad \widehat{\chi}_\tau(\lambda y) = \lambda\tau \, .$$

This induces the isomorphism

$$\chi : T_L\mathbb{P}_N \longrightarrow \mathrm{Hom}\,(L, \mathbb{C}^{N+1}/L) \, .$$

In terms of tangent vectors to paths, χ can be described as follows. Let $S \subset \mathbb{C}$ be an open set containing o, $\varrho : S \to \mathbb{C}^{N+1}$ a holomorphic path with $\varrho(o) = y$, and $\xi := \pi \circ \varrho$, then

$$\xi'(o) = d_y\pi(\varrho'(o)) \quad \text{and} \quad \chi_{\xi'(o)}(y) = \varrho'(o) + L \, .$$

This description of $T_L\mathbb{P}_N$, which only depends on $L \subset \mathbb{C}^{N+1}$, can be generalized to arbitrary Grassmannians [H, 16.1].

1.4.2 The Tangent Space of the Grassmannian

Returning to the projective notation, we consider

$$\mathbb{G}(k, N) := \mathrm{Gr}(k + 1, N + 1) \, ,$$

the variety of k-planes $E \subset \mathbb{P}_N$ or of $(k + 1)$-dimensional vector spaces $E \subset \mathbb{C}^{N+1}$. If $S \subset \mathbb{C}$ is open and

$$\xi : S \longrightarrow \mathbb{G}(k, N)\,, \quad \varrho : S \longrightarrow \mathbb{C}^{N+1}$$

are holomorphic paths, then ϱ is called a *moving vector* of ξ if $\varrho(s) \in \xi(s)$ for every $s \in S$. In the case $k = 0$ considered above, ϱ is just a *lifting* of ξ.

Theorem. *For $E \in \mathbb{G}(k, N)$, $\tau \in T_E\,\mathbb{G}(k, N)$ and $v \in E \subset \mathbb{C}^{N+1}$ we choose a domain $S \subset \mathbb{C}$ containing the origin o together with*

- *a holomorphic path $\xi : S \longrightarrow \mathbb{G}(k, N)$ such that $\xi(o) = E$ and $\xi'(o) = \tau$*
- *a moving vector $\varrho : S \longrightarrow \mathbb{C}^{N+1}$ of ξ such that $\varrho(o) = v$.*

Then the map

$$\chi : T_E\,\mathbb{G}(k, N) \longrightarrow \mathrm{Hom}\,(E, \mathbb{C}^{N+1}/E)\,, \quad \tau \longmapsto \chi_\tau\,, \quad \text{with } \chi_\tau(v) = \varrho'(o) + E \, ,$$

is well defined and an isomorphism of vector spaces.

In a short form the effect of χ may be written as

$$\chi_{\xi'(o)}(\varrho(o)) = \varrho'(o) + \xi(o) \, .$$

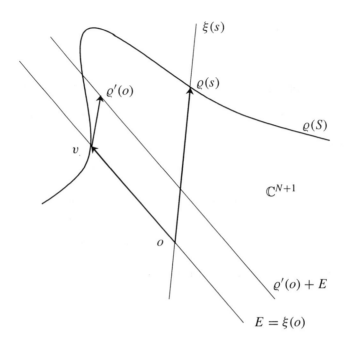

Proof. One advantage of the map χ is that it is canonical. Nevertheless, the easiest way to see its properties is by using coordinates. Since the Grassmannian is homogeneous, we may assume

$$E = \text{span}\{e_0, \dots, e_k\} \subset \mathbb{C}^{N+1} .$$

With the notation of 1.2.1, we put $I = \{0, \dots, k\}$ and consider the coordinate map

$$\Phi_I : M \longrightarrow U_I \subset \mathbb{G}(k, N) ,$$

where $M := M((N - k) \times (k + 1); \mathbb{C})$ and $E = \Phi_I(0)$. Every path $\xi : S \longrightarrow U_I$ can be expressed in these coordinates by functions $\xi_{ij}(s), k+1 \le i \le N, 0 \le j \le k$. Furthermore Φ_I induces an isomorphism

$$T_E\mathbb{G}(k, N) \longrightarrow M , \quad \xi'(o) \longmapsto (\xi'_{ij}(o)) \qquad (*)$$

if $\xi(o) = E$. This is the first part of χ.

For the second part, we use the isomorphism

$$M \longrightarrow \text{Hom}(E, \mathbb{C}^{N+1}/E) , \quad B \longmapsto \chi_B , \quad \text{with } \chi_B(e_j) = \sum_{i=k+1}^{N} b_{ij}e_i + E \qquad (**)$$

where $B = (b_{ij})$ for $k + 1 \le i \le N, 0 \le j \le k$.

We want to prove that the composite of the isomorphisms $(*)$ and $(**)$ gives the map χ as in the statement of the theorem for any choice of ξ and ϱ. The assumption $\varrho(s) \in \xi(s)$

together with the definition of Φ_I in 1.2.1 and CRAMER's rule in 1.1.3 imply that there are holomorphic functions $\varrho_0, \dots, \varrho_k$ on S such that

$$\varrho(s) = \sum_{j=0}^{k} \varrho_j(s) \left(e_j + \sum_{i=k+1}^{N} \xi_{ij}(s) e_i \right).$$

Now, $E = \Phi_I(0)$ implies $\xi_{ij}(o) = 0$; hence, $v = \varrho(o) = \sum_{j=0}^{k} \varrho_j(o) e_j$.

Differentiating this expression of $\varrho(s)$ and taking values at $s = o$ yields

$$\varrho'(o) = \sum_j \varrho_j'(o) \, e_j + \sum_{i,j} \varrho_j(o) \, \xi_{ij}'(o) \, e_i .$$

Since the first summand on the right is in E, we obtain

$$\varrho'(o) + E = \left(\sum_{i,j} \varrho_j(o) \, \xi_{ij}'(o) \, e_i \right) + E = \chi_{\xi_{(o)}'}(\varrho(o)) = \chi_\tau(v) . \qquad \square$$

1.5   Curves in Grassmannians

The simplest non-trivial subvarieties of Grassmannians are curves. According to [GH₁], they can be represented in a *normal form*. The topic of this section is to give an elementary proof of this result (see [P₁]).

1.5.1 The Drill

By a *curve* $B \subset \mathbb{G}(k, N) = \mathrm{Gr}(k + 1, N + 1)$, we mean a one-dimensional variety. It has a holomorphic desingularisation

$$\psi : S \longrightarrow B \subset \mathbb{G}(k, N),$$

where S denotes a compact Riemann surface and ψ is biholomorphic outside a finite set. For technical reasons, the parameterized form is often easier to handle.

A curve in the Grassmannian represents a one parameter family of vector spaces. There are many ways of selecting vectors in these spaces. If $U \subset S$ is open, then a holomorphic map

$$\varrho : U \longrightarrow \mathbb{C}^{N+1}$$

was called a *moving vector* of ψ on U if $\varrho \in \psi$, i.e. $\varrho(s) \in \psi(s)$ for every $s \in U$. A family

$$\varrho_0, \dots, \varrho_k : U \longrightarrow \mathbb{C}^{N+1}$$

of moving vectors of ψ on U is called a *local basis* if

$$(\varrho_0(s), \dots, \varrho_k(s)) \quad \text{is a basis of } \psi(s)$$

for every $s \in U$. By 1.1.3, every moving vector ϱ of ψ has a representation

$$\varrho = \sum_{i=0}^{k} \lambda_i \varrho_i$$

with holomorphic functions λ_i on U.

Remark. *Given a holomorphic curve $\psi : S \to \mathbb{G}(k, N)$ and a point $o \in S$. Then there is an open neighborhood $o \in U \subset S$ with a local basis*

$$\varrho_0, \dots, \varrho_k : U \longrightarrow \mathbb{C}^{N+1}$$

of ψ on U.

Proof. We know $\psi(o) \in U_I$ for some $I = \{i_0, \dots, i_k\}$, as in paragraph 1.2.1; hence, there is a holomorphic matrix $B^I(s)$ such that $\psi(s)$ is spanned by the columns of $B^I(s)$ for every $s \in U := \psi^{-1}(U_I)$. Consequently, we can define $\varrho_j(s)$ by the j-th column of $B^I(s)$ for $j = 0, \dots, k$. $\square$

Every holomorphic map $\varrho : U \to \mathbb{C}^{N+1}$ determines a new holomorphic map

$$\varrho' : U \longrightarrow \mathbb{C}^{N+1}, \quad s \longmapsto \frac{d\varrho}{ds}(s),$$

where s is a local coordinate on U around o. If ϱ is a moving vector for ψ, then this need not be the case for ϱ'. The variation or the "drill" of a curve ψ can be measured by the span of all ϱ'. The main technical tool is a *distinguished local basis*.

Main Lemma. *For every holomorphic curve*

$$\psi : S \longrightarrow \mathbb{G}(k, N)$$

there is a unique integer d with $0 \le d \le k + 1$ such that for every point $o \in S$ there is an open neighborhood U and moving vectors $\varrho_0, \dots, \varrho_k$ on U, such that for every $s \in U \setminus \{o\}$ the following holds:

1) $(\varrho_0(s), \dots, \varrho_k(s))$ is a basis of $\psi(s)$

2) $(\varrho_0(s), \dots, \varrho_k(s), \varrho_0'(s), \dots, \varrho_{d-1}'(s))$ is a basis of
   span $\{\varrho_0(s), \dots, \varrho_k(s), \varrho_0'(s), \dots, \varrho_k'(s)\}$

3) $\varrho_d'(s), \dots, \varrho_k'(s) \in$ span $\{\varrho_0(s), \dots, \varrho_k(s)\}$.

We call $d = d(\psi)$ the *drill* of ψ. The obvious bounds are $0 \le d(\psi) \le k + 1$ and $k + d(\psi) \le N$.

Proof. Take $o \in S$, consider a local basis $\sigma_0, \ldots, \sigma_k$ on some U as in the above remark, and define

$$n(s) := \text{rank } (\sigma_0(s), \ldots, \sigma_k(s), \sigma_0'(s), \ldots, \sigma_k'(s)).$$

If $n := \max\{n(s) : s \in U\}$, then $U' := \{s \in U : n(s) < n\}$ has only isolated points. An easy computation shows that n does not depend on the choice of the local basis, so we can define

$$d := n - k - 1.$$

Shrink U such that $n(s) = n$ for $s \in U \setminus \{o\}$. By relabeling, we can assume that property 2) is satisfied for $(\sigma_0, \ldots, \sigma_k, \sigma_0', \ldots, \sigma_{d-1}')$. Hence, for $i = d, \ldots, k$ we know that the functions

$$\sigma_0, \ldots, \sigma_k, \sigma_0', \ldots, \sigma_{d-1}', \sigma_i'$$

are linearly dependent on U. By using CRAMER's rule (1.1.3) we find holomorphic functions a_{ij}, b_{ij}, c_i on U such that $c_i(s) \neq 0$ for $s \neq o$ and

$$\sum_{j=0}^{k} a_{ij}\sigma_j + \sum_{j=0}^{d-1} b_{ij}\sigma_j' + c_i \sigma_i' = 0.$$

Now we define $\varrho_0 := \sigma_0, \ldots, \varrho_{d-1} := \sigma_{d-1}$ and

$$\varrho_i := \sum_{j=0}^{d-1} b_{ij}\sigma_j + c_i \sigma_i \quad \text{for } i = d, \ldots, k.$$

Then *1)* and *2)* follow from $c_i(s) \neq 0$ and *3)* follows from

$$\varrho_i' = \sum_{j=0}^{d-1}(b_{ij}'\sigma_j + b_{ij}\sigma_j') + c_i'\sigma_i + c_i \sigma_i'$$
$$= \sum_{j=0}^{d-1} b_{ij}'\sigma_j - \sum_{j=0}^{k} a_{ij}\sigma_j + c_i'\sigma_i.$$

$\square$

1.5.2 Derived Curves

We continue the investigation of the variation of a curve, $\psi : S \to \mathbb{G}(k, N)$, by looking at the possible values at o of the derivatives ϱ' of moving vectors ϱ in some neighborhood of $o \in S$. More precisely, we define

$$\psi_+(o) := \{v \in \mathbb{C}^{N+1} : v = \varrho'(o) \text{ for some } \varrho \in \psi\}$$

$$\psi_-(o) := \{v \in \mathbb{C}^{N+1} : v = \varrho(o) \text{ for some } \varrho \in \psi \text{ with } \varrho'(o) \in \psi(o)\}.$$

These are vector spaces with

$$\psi_-(o) \subset \psi(o) \subset \psi_+(o)$$

for every $o \in S$. The first inclusion is trivial. For the second inclusion, take $v \in \psi(o)$, a moving vector σ of ψ around o such that $\sigma(o) = v$, and a local coordinate s of S centered in o. If

$$\varrho := (s - o)\sigma, \quad \text{then } \varrho' = \sigma + (s - o)\sigma', \quad \text{hence } \varrho'(o) = \sigma(o) = v.$$

Now we show that these two constructions yield *derived curves*, which are classical in case $k = 0$ (1.5.4).

Proposition. *Given a compact Riemann surface S and a holomorphic curve*

$$\psi : S \longrightarrow \mathbb{G}(k, N)$$

of drill d, there exist holomorphic curves

$$\psi^{(1)} : S \longrightarrow \mathbb{G}(k + d, N), \quad \psi_{(1)} : S \longrightarrow \mathbb{G}(k - d, N)$$

such that $\psi^{(1)}(s) = \psi_+(s)$ and $\psi_{(1)}(s) = \psi_-(s)$ for almost every $s \in S$. These constructions are dual in the following sense:

$$\mathcal{D}(\psi_{(1)}) = (\mathcal{D}\psi)^{(1)} \quad and \quad \mathcal{D}(\psi^{(1)}) = (\mathcal{D}\psi)_{(1)}.$$

Proof. We take a point $o \in S$ and a neighborhood U with a distinguished local basis $\varrho_0, \ldots, \varrho_k$ as in the Main Lemma. On $U \setminus \{o\}$ we have the equalities

$$\psi_+ = \text{span}\{\varrho_0, \ldots, \varrho_k, \varrho'_0, \ldots, \varrho'_{d-1}\} \quad \text{and} \quad \psi_- = \text{span}\{\varrho_d, \ldots, \varrho_k\}.$$

The first one is obvious. For the second one "$\supset$" follows from property *3)*. Conversely, take $\varrho(s) \in \psi_-(s)$. Then

$$\varrho = \sum_{j=0}^{k} \alpha_j \varrho_j, \quad \varrho' = \sum_{j=0}^{k} (\alpha'_j \varrho_j + \alpha_j \varrho'_j) = \sum_{j=0}^{k} \beta_j \varrho_j + \sum_{j=0}^{d-1} \alpha_j \varrho'_j,$$

and $\varrho'(s) \in \psi(s)$ implies $\alpha_0(s) = \ldots = \alpha_{d-1}(s) = 0$.

This argument also shows that

$$\psi_-(s) = \{\varrho(s) : \varrho \in \psi \quad \text{with } \varrho' \in \psi\}$$

for almost every $s \in S$.

Let us compare these different descriptions of the two vector spaces. The definitions of ψ_+ and ψ_- are independent of a basis; the description by a moving base is only local but shows the holomorphic dependence. The two descriptions together yield a finite set $S' \subset S$ and holomorphic maps

$$\psi^{(1)} : S \setminus S' \longrightarrow \mathbb{G}(k + d, N) \quad \text{and} \quad \psi_{(1)} : S \setminus S' \longrightarrow \mathbb{G}(k - d, N)$$

such that $\psi^{(1)}(s) = \psi_+(s)$ and $\psi_{(1)} = \psi_-(s)$ for $s \in S \setminus S'$. By 1.1.4, there are unique holomorphic extensions to S.

For the proof of the duality result, we first note that there are inclusions

$$\psi_{(1)} \quad \subset \quad \psi \quad \subset \quad \psi^{(1)},$$

$$\mathcal{D}(\psi_{(1)}) \supset \mathcal{D}\psi \supset \mathcal{D}(\psi^{(1)})$$

$$\|$$

$$(\mathcal{D}\psi)^{(1)} \supset \mathcal{D}\psi \supset (\mathcal{D}\psi)_{(1)}.$$

It suffices to prove the equality $\mathcal{D}(\psi_{(1)}) = (\mathcal{D}\psi)^{(1)}$, and we may use the inclusion

$$\mathcal{D}((\mathcal{D}\psi)^{(1)}) \subset \mathcal{D}(\mathcal{D}\psi) = \psi.$$

We denote the canonical symmetric bilinear form on $\mathbb{C}^{N+1}$ by

$$\langle v, w \rangle = \sum_{j=0}^{N} v_j w_j.$$

Given moving vectors $\varrho, \sigma : U \to \mathbb{C}^{N+1}$ of ψ and $\mathcal{D}\psi$, we have $\langle \varrho, \sigma \rangle = 0$ on U and hence

$$0 = \langle \varrho, \sigma \rangle' = \langle \varrho', \sigma \rangle + \langle \varrho, \sigma' \rangle.$$

For a moving vector ϱ of ψ and $s \in U$, this implies

$$\begin{aligned}
&\langle \varrho(s), \sigma'(s) \rangle &&= 0 &&\text{for all } \sigma \in \mathcal{D}\psi \\
\Longleftrightarrow\ &\langle \varrho'(s), \sigma(s) \rangle &&= 0 &&\text{for all } \sigma \in \mathcal{D}\psi \\
\Longleftrightarrow\ &\langle \varrho'(s), w \rangle &&= 0 &&\text{for all } w \in (\mathcal{D}\psi)(s) \\
\Longleftrightarrow\ &\varrho'(s) \in \mathcal{D}(\mathcal{D}\psi)(s) = \psi(s).
\end{aligned}$$

For almost every point $s \in U$, we have

$$(\mathcal{D}\psi)^{(1)}(s) = \{\sigma'(s) : \sigma \in \mathcal{D}\psi\}.$$

By using all the above remarks, we finally obtain for almost every $s \in S$

$$\mathcal{D}((\mathcal{D}\psi)^{(1)})(s) = \{v \in \mathbb{C}^{N+1} : \langle v, \sigma'(s) \rangle = 0 \quad \text{for all } \sigma \in \mathcal{D}\psi\}$$

$$= \{\varrho(s) : \varrho \in \psi \quad \text{such that } \langle \varrho(s), \sigma'(s) \rangle = 0 \quad \text{for all } \sigma \in \mathcal{D}\psi\}$$

$$= \{\varrho(s) : \varrho'(s) \in \psi(s)\} = \psi_{(1)}(s). \qquad \square$$

For later use we note some additional properties of the derived curves.

Remark. *With the notations as before, we have:*

1) $\psi = \psi^{(1)}$ *iff* ψ *is constant,*

2) $(\psi_{(1)})^{(1)} \subset \psi$, *in particular* $d(\psi_{(1)}) \leq d(\psi)$.

Proof. 1) Assume ψ is constant and take a basis $(v_0, \dots, v_k)$ of $V = \psi(s)$. If ϱ is a moving vector on $U \subset S$, then there are holomorphic functions λ_i on U such that

$$\varrho = \sum_{i=0}^{k} \lambda_i v_i , \quad \text{hence} \quad \varrho' = \sum_{i=0}^{k} \lambda_i' v_i \in V ,$$

and this implies $\psi^{(1)} \subset \psi$.

Conversely, if $\psi^{(1)} = \psi$ then for every local basis $(\varrho_0, \dots, \varrho_k)$ on U we have holomorphic functions λ_i^j on U such that

$$\varrho_j' = \sum_{i=0}^{k} \lambda_i^j \varrho_i \quad \text{for } j = 0, \dots, k .$$

We consider the holomorphic maps

$$\varrho : U \longrightarrow \bigwedge^{k+1} \mathbb{C}^{N+1}, \quad s \longmapsto \varrho_0(s) \wedge \dots \wedge \varrho_k(s), \quad \text{and}$$

$$\varrho' : U \longrightarrow \bigwedge^{k+1} \mathbb{C}^{N+1}, \quad s \longmapsto \sum_{i=0}^{k} \varrho_0(s) \wedge \dots \wedge \varrho_i'(s) \wedge \dots \wedge \varrho_k(s)$$

$$= \left(\sum_{i=0}^{k} \lambda_i^i(s) \right) \varrho_0(s) \wedge \dots \wedge \varrho_k(s) .$$

This implies $\varrho'(s) = \alpha(s)\varrho(s)$ with α holomorphic; hence, by Lemma 1.1.5, there is some $\omega = v_0 \wedge \dots \wedge v_k$ and a holomorphic β on U such that

$$\varrho(s) = \beta(s)\omega .$$

Consequently, ψ is constant on U, hence on S.

2) This is an immediate consequence of the relations

$$\psi_{(1)} = \text{span} \{\varrho_d, \dots, \varrho_k\} \quad \text{and} \quad \varrho_d', \dots, \varrho_k' \in \text{span} \{\varrho_0, \dots, \varrho_k\}$$

which are valid almost everywhere for a distinguished local basis. $\square$

Obviously, the construction of derived curves given in the above proposition can be iterated: if $\psi : S \to \mathbb{G}(k, N)$ is holomorphic, we define

$$\psi_{(0)} := \psi =: \psi^{(0)}, \quad \psi_{(r+1)} := (\psi_{(r)})_{(1)}, \quad \psi^{(r+1)} := (\psi^{(r)})^{(1)} .$$

The geometric effect of this iteration will be investigated in 1.5.4.

1.5.3 Sums and Intersections

A moving vector defined globally on a compact Riemann surface S is constant. An adequate global concept is the following: given two holomorphic curves

$$\varphi : S \longrightarrow \mathbb{G}(0, N) \quad \text{and} \quad \psi : S \longrightarrow \mathbb{G}(k, N),$$

then φ is called a *directrix* of ψ, if $\varphi \subset \psi$, i.e. $\varphi(s) \subset \psi(s)$ for every $s \in S$. We will construct an analogue of a local basis for ψ with these directrices.

We first consider a more general situation. For holomorphic curves

$$\varphi : S \longrightarrow \mathbb{G}(k, N) = \mathrm{Gr}(k+1, N+1) \quad \text{and} \quad \psi : S \longrightarrow \mathbb{G}(l, N) = \mathrm{Gr}(l+1, N+1)$$

we want to construct new holomorphic curves

$$\varphi + \psi : S \longrightarrow \mathbb{G}(m-1, N) = \mathrm{Gr}(m, N+1)$$

and

$$\varphi \cap \psi : S \longrightarrow \mathbb{G}(n-1, N) = \mathrm{Gr}(n, N+1).$$

For this purpose, we define

$$m(s) := \dim(\varphi(s) + \psi(s)) \quad \text{and} \quad n(s) := \dim(\varphi(s) \cap \psi(s)),$$

$$m := \max\{m(s) : s \in S\} \quad \text{and} \quad n := \min\{n(s) : s \in S\},$$

$$S' := \{s \in S : m(s) < m\} = \{s \in S : n(s) > n\}.$$

These two exceptional sets are equal, since $m(s) + n(s) = k + l + 2$ for every $s \in S$, and S' is finite. We consider the maps

$$\varphi + \psi : S \setminus S' \longrightarrow \mathbb{G}(m-1, N), \quad s \longmapsto \varphi(s) + \psi(s), \quad \text{and}$$
$$\varphi \cap \psi : S \setminus S' \longrightarrow \mathbb{G}(n-1, N), \quad s \longmapsto \varphi(s) \cap \psi(s).$$

It is easy to see that $\varphi + \psi$ is holomorphic: locally there are local bases of φ and ψ, and one can select $U \subset S$ open and

$$\varrho_1, \ldots, \varrho_m : U \longrightarrow \mathbb{C}^{N+1}$$

such that $(\varrho_1(s), \ldots, \varrho_m(s))$ is a basis of $\varphi(s) + \psi(s)$ for every $s \in U$. By 1.1.4, the map $\varphi + \psi$ has a unique holomorphic extension to S.

The holomorphic curve $\varphi \cap \psi$ can be derived from a holomorphic sum by duality, since

$$\mathcal{D}(\varphi \cap \psi) = \mathcal{D}\varphi + \mathcal{D}\psi.$$

The sum $\varphi + \psi$ is called *direct* if $m = k + l + 2$, which is equivalent to $n = 0$; we write

$$\varphi + \psi = \varphi \oplus \psi.$$

Using this concept of direct sums, we can show the existence of sufficiently many directrices.

Lemma. *Given a compact Riemann surface S and two curves*

$$\varphi : S \longrightarrow \mathbb{G}(k, N), \quad \psi : S \longrightarrow \mathbb{G}(l, N) \quad with \; \varphi \subset \psi \, ,$$

there exist directrices $\varphi_1, \dots , \varphi_{l-k} : S \to \mathbb{G}(0, N)$ *of* ψ *such that*

$$\psi = \varphi \oplus \varphi_1 \oplus \dots \oplus \varphi_{l-k} \, .$$

Proof. Take an arbitrary point $o \in S$ and a base

$$(v_0, \dots , v_k, w_1, \dots , w_{l-k}, u_1, \dots , u_{N-l}) \quad \text{of} \quad \mathbb{C}^{N+1}$$

such that the first $k + 1$ vectors are a base of $\varphi(o)$ and the first $l + 1$ vectors are a base of $\psi(o)$. Define

$$W_i := \mathrm{span}\,\{w_i, u_1, \dots , u_{N-l}\} \subset \mathbb{C}^{N+1} \quad \text{for } i = 1, \dots , l - k \, ,$$

and denote by ψ_i the constant curve with $\psi_i(s) = W_i$ for all s. We claim that $\varphi_i := \psi \cap \psi_i$ is a directrix, i.e.

$$\varphi_i : S \longrightarrow \mathbb{G}(0, N) \, .$$

Therefore we have to show $\dim(\psi(s) \cap W_i) = 1$ for almost all s. For $s = o$, this is true since by construction

$$\psi(o) \cap W_i = \mathbb{C}\,w_i \, .$$

For every s, we know $\dim \psi(s) + \dim W_i = N + 2$, hence

$$\dim(\psi(s) \cap W_i) \geq 1 \, ,$$

and the assertion follows from the semicontinuity of the dimension. By definition of φ_i, the equation

$$\psi(s) = \varphi(s) \oplus \varphi_1(s) \oplus \dots \oplus \varphi_{l-k}(s)$$

holds for $s = o$, and again by semicontinuity it follows for almost all $s \in S$. $\qquad \square$

The following result is very easy to prove:

Remark. *Assume* $\psi : S \to \mathbb{G}(k, N)$ *has a representation*

$$\psi = \psi_1 \oplus \dots \oplus \psi_e \, .$$

Then

$$\psi_{(1)} \quad \supset \quad (\psi_1)_{(1)} \oplus \dots \oplus (\psi_e)_{(1)} \, , \quad and$$

$$\psi^{(1)} \quad = \quad \psi_1^{(1)} + \dots + \psi_e^{(1)} \, .$$

In general, the inclusion is proper, and the second sum is not direct.

1.5.4 Associated Curves and Curves with Prescribed Drill

The construction of derived curves of 1.5.2 is classical in the case $k = 0$, i.e. in the case of a curve

$$\varphi : S \longrightarrow \mathbb{G}(0, N) = \mathrm{Gr}(1, N + 1) = \mathbb{P}_N(\mathbb{C}).$$

We recall some basic facts: if $U \subset S$ is a coordinate neighborhood and

$$\varrho : U \longrightarrow \mathbb{C}^{N+1} \setminus \{0\}$$

is a moving vector (or *local lifting*) of φ, i.e. $\varrho \in \varphi$, then we consider the sequence of derivatives

$$\varrho, \varrho', \dots, \varrho^{(r)}, \dots : U \longrightarrow \mathbb{C}^{N+1}$$

with respect to the coordinate on U. It is easy to see that the number

$$\mathrm{rank}\, \varphi := \max\{r \in \mathbb{N} : \varrho(s), \varrho'(s), \dots, \varrho^{(r)}(s) \in \mathbb{C}^{N+1} \text{ are linearly independent}$$
$$\text{for some } \varrho \in \varphi \text{ and some } s \in U\}$$

only depends on φ, $0 \leq \mathrm{rank}\, \varphi \leq N$, and that $\mathrm{rank}\, \varphi$ is the minimal dimension of a linear space E such that

$$\varphi(S) \subset E \subset \mathbb{P}_N(\mathbb{C}).$$

In particular, φ is called *nondegenerate* if $\mathrm{rank}\, \varphi = N$.

For $0 \leq k \leq \mathrm{rank}\, \varphi$, there is the k-th *associated curve*

$$\varphi^{(k)} : S \longrightarrow \mathbb{G}(k, N) \subset \mathbb{P}(\bigwedge^{k+1} \mathbb{C}^{N+1}),$$

which is defined locally for almost all $s \in U$ by

$$\varphi^{(k)}(s) = \mathbb{C}(\varrho(s) \wedge \varrho'(s) \wedge \dots \wedge \varrho^{(k)}(s)).$$

The use of the same notation as for derived curves (1.5.2) is justified by the

Remark. *Take a curve* $\varphi : S \to \mathbb{G}(0, N)$ *and define* $\psi := \varphi^{(k)}$. *Then*

$$\psi^{(1)} = \varphi^{(k+1)} \quad \textit{for } 0 \leq k < \mathrm{rank}\, \varphi \qquad \textit{and}$$
$$\psi_{(1)} = \varphi^{(k-1)} \quad \textit{for } 0 < k \leq \mathrm{rank}\, \varphi.$$

Now we compute the drill $d(\varphi^{(k)})$ of the associated curves of a curve

$$\varphi : S \longrightarrow \mathbb{P}_N(\mathbb{C})$$

with a local lifting ϱ. For $0 \leq k \leq \mathrm{rank}\, \varphi$ a local basis of $\varphi^{(k)}$ is given by

$$(\varrho_0, \dots, \varrho_k) = (\varrho, \varrho', \dots, \varrho^{(k)});$$

hence, we have

$$(\varrho_0, \ldots, \varrho_k, \varrho_0', \ldots, \varrho_k') = (\varrho, \varrho', \ldots, \varrho^{(k)}, \varrho', \varrho'', \ldots, \varrho^{(k+1)}) \,.$$

This immediately implies

$$d(\varphi^{(k)}) = \begin{cases} 1 & \text{for} \ \ 0 \le k < \text{rank} \, \varphi \,, \\ 0 & \text{for} \ \ k = \text{rank} \, \varphi \,. \end{cases}$$

If we have curves $\varphi_i : S \to \mathbb{P}_N(\mathbb{C})$ for $i = 1, \ldots, m$ and integers k_i with $0 \le k_i \le \text{rank} \, \varphi_i$, we may consider the sum

$$\varphi_1^{(k_1)} + \ldots + \varphi_m^{(k_m)} : S \to \mathbb{G}(k, N) \,,$$

where $k \le k_1 + \ldots + k_m + m - 1$, with equality in case the sum is direct. Then it is clear that

$$d(\varphi_1^{(k_1)} + \ldots + \varphi_m^{(k_m)}) \le m$$

and

$$d(\varphi_1^{(k_1)} \oplus \ldots \oplus \varphi_m^{(k_m)}) = m$$

in case $\varphi_1^{(k_1+1)} + \ldots + \varphi_m^{(k_m+1)}$ is direct.

With these methods we obtain the following

Proposition. *For given integers d, k, N satisfying the obvious restrictions*

$$0 \le d \le k+1 \quad \text{and} \quad k+d \le N \,,$$

and any Riemann surface S, there exists a curve

$$\psi : S \longrightarrow \mathbb{G}(k, N)$$

of drill d.

Proof. We start with any choice of integers $k_1, \ldots, k_d$ such that

$$k_1 + \ldots + k_d + d = k + 1 \,.$$

Then

$$N \ge k + d = k_1 + \ldots + k_d + 2d - 1 \,.$$

It suffices to show that there are curves

$$\psi_i : S \longrightarrow \mathbb{P}_N \quad \text{with rank} \, \psi_i \ge k_i + 1$$

for $i = 1, \ldots, d$ such that the sum

$$\psi_1^{(k_1+1)} + \ldots + \psi_d^{(k_d+1)}$$

is direct. Then we can define

$$\psi := \psi_1^{(k_1)} \oplus \ldots \oplus \psi_d^{(k_d)},$$

and the above computation shows that ψ has drill d.

For $i = 1, \ldots, d$ we choose a nondegenerate curve

$$\varphi_i : S \longrightarrow \mathbb{P}_{k_i+1}.$$

For example, if $S = \mathbb{P}_1$, we may take the rational normal curve.

If $n = k + d$, there is a decomposition

$$\mathbb{C}^{n+1} = \mathbb{C}^{k_1+2} \oplus \ldots \oplus \mathbb{C}^{k_d+2},$$

which yields inclusion maps

$$\varepsilon_i : \mathbb{P}_{k_i+1} \longrightarrow \mathbb{P}_n \subset \mathbb{P}_N.$$

So we may take $\psi_i := \varepsilon_i \circ \varphi_i$. $\square$

1.5.5 Normal Form

The main result of this section is that every curve in a Grassmannian can be decomposed in the same way as we have constructed curves of given drill in 1.5.4.

Normal Form Lemma. *Take a compact Riemann surface S and a holomorphic map*

$$\psi : S \longrightarrow \mathbb{G}(k, N)$$

of drill d with $0 \le d \le k+1$. Then there exist integers $a_1, \ldots, a_d$ with $a_1 \ge \ldots \ge a_d \ge 0$, holomorphic curves

$$\varphi_i : S \longrightarrow \mathbb{G}(0, N), \quad i = 1, \ldots, d,$$

with $\operatorname{rank} \varphi_i \ge a_i + 1$ and a linear space $L \subset \mathbb{P}_N$ such that

$$\psi = L \oplus \varphi_1^{(a_1)} \oplus \ldots \oplus \varphi_d^{(a_d)} \qquad and$$

$$\psi^{(1)} = L \oplus \varphi_1^{(a_1+1)} \oplus \ldots \oplus \varphi_d^{(a_d+1)}.$$

In particular, $L = \bigcap_{s \in S} \psi(s)$ and $k = \dim L + a_1 + \ldots + a_d + d$.

Moreover $a_1, \ldots, a_d$ and L are uniquely determined. In case $L = \emptyset$ and additionally $d = 1$ or $a_1 > a_2$ the curve φ_1 is uniquely determined.

Proof. We proceed by induction on k. For $k = -1$, $k = 0$, or ψ constant there is nothing to prove.

Consequently, we may apply the induction hypothesis to $\psi_{(1)}$ as follows.

If $e := d(\psi_{(1)})$, there are $\varphi_1, \ldots, \varphi_e$, $b_1 \geq \ldots \geq b_e \geq 0$ and $L \subset \mathbb{P}_N$ such that

$$\psi_{(1)} = L \oplus \varphi_1^{(b_1)} \oplus \ldots \oplus \varphi_e^{(b_e)} \qquad \text{and}$$

$$(\psi_{(1)})^{(1)} = L \oplus \varphi_1^{(b_1+1)} \oplus \ldots \oplus \varphi_e^{(b_e+1)} .$$

For every $s \in S$, we have

$$\dim \psi_{(1)}(s) \quad = \quad \dim L + b_1 + \ldots + b_e + e \quad = \quad k - d ,$$

$$\dim(\psi_{(1)})^{(1)}(s) \quad = \quad \dim L + b_1 + \ldots + b_e + 2e \quad = \quad k - d + e .$$

By Remark 1.5.2, we know $e \leq d$, and by Lemma 1.5.3, we find $\varphi_{e+1}, \ldots, \varphi_d$ such that

$$\psi = L \oplus \varphi_1^{(b_1+1)} \oplus \ldots \oplus \varphi_e^{(b_e+1)} \oplus \varphi_{e+1} \oplus \ldots \oplus \varphi_d .$$

This yields $a_1 = b_1 + 1, \ldots, a_e = b_e + 1, a_{e+1} = \ldots = a_d = 0$. Obviously,

$$\psi^{(1)} = L + \varphi_1^{(a_1+1)} + \ldots + \varphi_e^{(a_e+1)} + \varphi_{e+1}^{(1)} + \ldots + \varphi_d^{(1)} .$$

Since $\dim \psi^{(1)}(s) = \dim \psi(s) + d$ for almost all s, this sum is direct.

Further, the $a_1, \ldots, a_d$ are unique, because they can be computed from $b_1, \ldots, b_e$, which are unique by induction hypothesis. An analogous argument ensures the eventual uniqueness of φ_1. $\qquad\square$

Chapter 2

Ruled Varieties

In this chapter, we study algebraic subvarieties of affine or projective spaces which are unions of linear subspaces.

2.1 Incidence Varieties and Duality

Given a subvariety $X \subset \mathbb{P}_N$, we can try to find all linear subspaces of some dimension, $k \leq \dim X$, contained in X. The result is the Fano variety $F_k(X) \subset \mathbb{G}(k, N)$. Conversely, we can start with any subvariety $B \subset \mathbb{G}(k, N)$ and consider the union $X \subset \mathbb{P}_N$ of all linear subspaces $E \in B$. We handle the second question first.

2.1.1 Unions of Linear Varieties

As always, we identify $\mathbb{G}(k, N) = \mathrm{Gr}(k + 1, N + 1)$, the Grassmannians of k-dimensional linear subspaces of $\mathbb{P}_N$ and of $(k + 1)$-dimensional vector subspaces of $\mathbb{C}^{N+1}$. Via the Plücker embedding, we have

$$\mathbb{G}(k, N) \subset \mathbb{P}_m \quad \text{with } m = \binom{N + 1}{k + 1} - 1,$$

and by using the fact that incidences in $\mathbb{P}_N$ can be expressed in terms of Plücker coordinates, we obtained in 1.3.2 the simple but fundamental

Lemma. *For* $0 \leq k \leq N$, *the incidence variety*

$$I_{k,N} := \mathbb{F}(0, k, N) = \{(x, E) \in \mathbb{P}_N \times \mathbb{G}(k, N) : x \in E\}$$

is algebraic in $\mathbb{P}_N \times \mathbb{P}_m$.

By using standard results in algebraic geometry, this implies the

Theorem. *If* $B \subset \mathbb{G}(k, N)$ *is algebraic, then*

$$X := \bigcup_{E \in B} E \subset \mathbb{P}_N$$

is algebraic. If B *is irreducible, then* X *is irreducible and* $\dim X \leq \dim B + k$.

Proof. Consider the subvariety $I_B := I_{k,N} \cap (\mathbb{P}_N \times B)$ with the canonical projections

$$
\begin{array}{ccc}
I_B & \xrightarrow{\ \xi\ } & B \\
\pi \downarrow & & \\
\mathbb{P}_N & &
\end{array}
$$

Obviously $\pi(I_B) = X$; hence, X is algebraic by the Regular Mapping Theorem (1.1.1). The map ξ is a bundle with fiber $\mathbb{P}_k$, so if B is irreducible, the same is true for I_B (1.1.1, Irreducibility Theorem) and also for $\pi(I_B)$. The Dimension Theorem for Regular Maps implies

$$
\dim X \leq \dim I_B = \dim B + k \,. \qquad \qquad \square
$$

Of course this last inequality may be strict; for example, take a plane $X \subset \mathbb{P}_3$ and

$$
B := \{L \in \mathbb{G}(1, 3) : L \subset X\} \,.
$$

In this case, $\dim B = \dim \mathbb{G}(1, 2) = 2$, $\dim I_B = 3$.

2.1.2 Fano Varieties

For $X \subset \mathbb{P}_N$, we define the k-th *Fano variety* of X by

$$
F_k(X) := \{E \in \mathbb{G}(k, N) : E \subset X\} \,.
$$

Proposition. *If $X \subset \mathbb{P}_N$ is algebraic, then $F_k(X) \subset \mathbb{G}(k, N)$ is algebraic for $0 \leq k \leq N$.*

Proof. If $X = X_1 \cap \ldots \cap X_r$, then obviously

$$
F_k(X) = F_k(X_1) \cap \ldots \cap F_k(X_r) \,.
$$

Hence it suffices to consider the case of a hypersurface. Take $X = V(F)$, where $F \in \mathbb{C}[x_0, \ldots, x_N]$ is homogeneous of degree d. Let $E = \mathbb{P}(\mathrm{span}\{v_0, \ldots, v_k\})$ be a k-dimensional linear subspace, then

$$
\mathbb{P}_k \longrightarrow \mathbb{P}_N, \quad (\lambda_0 : \ldots : \lambda_k) \longmapsto \mathbb{C}\left(\sum_{i=0}^{k} \lambda_i v_i\right),
$$

is a parameterization of E. If $v_i = {}^t(v_{i0}, \ldots, v_{iN})$, then E is contained in $V(F)$ iff the polynomial

$$
G_v(\lambda_i) := F\left(\sum_{i=0}^{k} \lambda_i v_i\right) \in (\mathbb{C}[v_{ij}])[\lambda_i]
$$

vanishes. The vanishing of the coefficients of G_v, which are homogeneous polynomials of degree d in the variables v_{ij}, has to be expressed in terms of Plücker coordinates of E. It is enough to do this in every coordinate neighborhood of the Grassmannian.

To simplify the notation, take $U_I \subset \mathbb{G}(k, N)$ with $I = \{0, \dots , k\}$. Then if $E \in U_I$, it can be spanned by the columns of the matrix

$$A = \begin{pmatrix} 1 & & & 0 \\ & \ddots & & \\ 0 & & 1 & \\ v_{0,k+1} & \cdots & v_{k,k+1} \\ \vdots & & \vdots \\ v_{0,N} & \cdots & v_{k,N} \end{pmatrix}.$$

With this convention, it suffices to show that the remaining variables v_{ij} with

$$0 \le i \le k, \quad k+1 \le j \le N,$$

are determined by the Plücker coordinates. Take

$$J := \{0, \dots , i-1, i+1, \dots , k, j\} \subset \{0, \dots , N\};$$

then, obviously, $p_J(E) = \det A_J = \pm v_{ij}$. $\qquad\qquad\qquad\qquad \square$

In general, the dimension of the Fano variety is very low. For instance, if $X \subset \mathbb{P}_3$ is a smooth surface of degree d, then it is well known that $F_1(X) \subset \mathbb{G}(1, 3) \subset \mathbb{P}_5$ is of the following form (see [Mu] or [Sh]):

- two disjoint rational curves for $d = 2$,
- 27 points for $d = 3$,
- empty for $d \ge 4$.

2.1.3 Joins

Given two varieties $X, Y \subset \mathbb{P}_N$, we recall the construction of their *join* $X \# Y$: if $\overline{xy}$ denotes the line joining two different points $x, y \in \mathbb{P}_N$, then

$$X \# Y := \text{closure of } \Big(\bigcup_{\substack{(x,y) \in X \times Y \\ x \ne y}} \overline{xy} \Big) \subset \mathbb{P}_N.$$

The following construction will make precise what the closure means, and shows that the result is algebraic.

There is a rational *join map*

$$J : \mathbb{P}_N \times \mathbb{P}_N \dashrightarrow \mathbb{G}(1, N), \quad J(x, y) = \overline{xy} \quad \text{for } x \ne y,$$

with indeterminacy along the diagonal $\Delta \subset \mathbb{P}_N \times \mathbb{P}_N$. Except for the case where $X = Y$ is one point, there is a restriction

$$J' : X \times Y \dashrightarrow \mathbb{G}(1, N)$$

of J, and we have an image $B := J'(X \times Y)$. By 2.1.1 the set

$$X \# Y := \bigcup_{L \in B} L \subset \mathbb{P}_N$$

is a variety, and

$$\dim(X \# Y) \leq \dim B + 1 \leq \dim X + \dim Y + 1 .$$

The number on the right is called the *expected dimension* of the join.

In case $X \cap Y = \emptyset$ the map J' is regular. If, in addition, the expected dimension is $\leq N$, then it is equal to $\dim(X \# Y)$ [H, 11.37].

In case $x \in X \cap Y$ it is a delicate question how many of the lines passing x lie in B [H, 15.15].

2.1.4 Conormal Bundle and Dual Variety

There are two kinds of slightly different projective spaces which are "dual" to $\mathbb{P}_N = \mathbb{G}(0, N) = \mathrm{Gr}(1, N + 1)$: the space of hyperplanes,

$$\mathbb{P}_N^\vee = \mathbb{G}(N - 1, N) = \mathrm{Gr}(N, N + 1) = \{H \subset \mathbb{C}^{N+1} \quad : H \text{ hyperplane}\} , \text{ and}$$
$$\mathbb{P}_N^* = \mathbb{G}^*(0, N) \qquad = \mathrm{Gr}^*(1, N + 1) = \{L \subset (\mathbb{C}^{N+1})^* : L \text{ line}\} ,$$

the space of linear forms modulo $\mathbb{C}^*$. They are isomorphic by duality (1.3.3)

$$\mathcal{D} = \mathcal{D}_{N-1,N} : \mathbb{G}(N - 1, N) \longrightarrow \mathbb{G}^*(0, N) ,$$

relating every hyperplane to its linear equations.

For a smooth hypersurface $X = V(F) \subset \mathbb{P}_N$, there is a regular map

$$X \xrightarrow{\delta} \mathbb{P}_N^*, \quad p \longmapsto \left(\frac{\partial F}{\partial x_0}(p) : \ldots : \frac{\partial F}{\partial x_N}(p) \right) ,$$

and the image $X^* := \delta(X) \subset \mathbb{P}_N^*$ is called the *dual variety*, which is essentially the same as

$$X^\vee := \mathcal{D}(\delta(X)) \subset \mathbb{P}_N^\vee ,$$

the *variety of tangent hyperplanes* of X. This construction can be generalized and analyzed.

The next step is the case of a general, not necessarily smooth, irreducible hypersurface $X = V(F) \subset \mathbb{P}_N$. Instead of a regular map, we obtain a rational map

$$\delta : X \dashrightarrow \mathbb{P}_N^*$$

in the following way. Consider the algebraic set

$$\Gamma_X := \left\{ (x; y) \in X \times \mathbb{P}_N^* : y_i \frac{\partial F}{\partial x_j}(x) - y_j \frac{\partial F}{\partial x_i}(x) = 0 \quad \text{for } i, j = 0, \dots, N \right\}.$$

This condition is equivalent to the linear dependence of y and $\mathrm{grad}_x F$, or

$$\mathrm{rank} \begin{pmatrix} y_0 & \cdots & y_N \\ \frac{\partial F}{\partial x_0}(x) & \cdots & \frac{\partial F}{\partial x_N}(x) \end{pmatrix} \le 1.$$

The projection $\pi : \Gamma_X \to X$ is biholomorphic outside the singular locus $\mathrm{Sing}\, X$. Hence, there is a unique irreducible component $\Gamma_\delta \subset \Gamma_X$ which is not contained in $\pi^{-1}(\mathrm{Sing}\, X)$. Now δ is defined as the rational map with graph Γ_δ, and the *dual variety* is

$$X^* := \delta(X) := \xi(\Gamma_\delta),$$

where $\xi : \Gamma_\delta \to \mathbb{P}_N^*$ denotes the projection map. Using duality, we obtain the rational map

$$\delta^\vee = \mathcal{D} \circ \delta : X \dashrightarrow \mathbb{P}_N^\vee, \quad \text{and} \quad X^\vee := \delta^\vee(X).$$

Geometrically, the image of a smooth point $p \in X$ under $\delta^\vee$ is $\mathbb{T}_p X$, the tangent hyperplane of X at p (1.1.1). The image $\delta^\vee(p)$ of $p \in \mathrm{Sing}\, X$ consists of all hyperplanes which are limits of tangent hyperplanes to smooth points near p. Formally, this limit procedure is established by the closure of Γ_δ in Γ_X.

Finally, for an irreducible variety $X \subset \mathbb{P}_N$ of any dimension n we want to define

$$X^\vee := \text{closure of } \{H \in \mathbb{P}_N^\vee : H \supset \mathbb{T}_p X \text{ for some smooth } p \in X\}.$$

This can be done in a more precise way. For $X = V(F_1, \dots, F_m)$, we also consider the affine cone $\widehat{X} \subset \mathbb{C}^{N+1}$ defined by the same polynomials. Let $p \in X$ be a smooth point and $q \in \widehat{X} \backslash \{0\}$ is lying over p, then $\widehat{X}$ is smooth in q and the tangent vector space is given by

$$T_q \widehat{X} = \left\{ x \in \mathbb{C}^{N+1} : \langle x, \mathrm{grad}_q F_k \rangle = 0 \quad \text{for } k = 1, \dots, m \right\}.$$

Recall the meaning of the bracket notation: for $x \in \mathbb{C}^{N+1}$ and $y \in (\mathbb{C}^{N+1})^*$, we write

$$\langle x, y \rangle = y(x).$$

For the dual space, we have

$$(T_q \widehat{X})^* = \mathrm{span}\, \{\mathrm{grad}_q F_1, \dots, \mathrm{grad}_q F_m\} \subset (\mathbb{C}^{N+1})^*.$$

We consider vectors in $(\mathbb{C}^{N+1})^*$ as rows and form the matrix

$$A(x, y) := \begin{pmatrix} y \\ \mathrm{grad}_x F_1 \\ \vdots \\ \mathrm{grad}_x F_m \end{pmatrix}$$

for $x \in \mathbb{C}^{N+1}$ and $y \in (\mathbb{C}^{N+1})^*$. Since the rank is determined by minors, the set

$$\widehat{\Gamma}_X := \{(x, y) \in \widehat{X} \times (\mathbb{C}^{N+1})^* : \operatorname{rank} A(x, y) \leq N - n\}$$

is algebraic. Furthermore, for a smooth point $x \in \widehat{X}$, we know

$$\operatorname{rank} A(x, y) \leq N - n \Longleftrightarrow y \in (T_x\widehat{X})^* \Longleftrightarrow \langle T_x\widehat{X}, y \rangle = 0.$$

Passing to the projective spaces, we obtain the

Lemma. *For a variety $X \subset \mathbb{P}_N$ there is an algebraic set $\Gamma_X \subset X \times \mathbb{P}_N^*$ with the following property: if $\pi : \Gamma_X \to X$ denotes the canonical projection, then*

$$\Gamma_{X,p} := \pi^{-1}(p) = \{(p, y) \in \{p\} \times \mathbb{P}_N^* : \langle \mathbb{T}_p X, y \rangle = 0\}$$

for every smooth point $p \in X$.

Now we follow the standard closing procedure. Outside the singular locus of X, the map π is a bundle with fiber $\mathbb{P}_{N-n-1}$. The irreducible component of Γ_X containing this bundle space is denoted by C_X, and called the *conormal bundle* of X. We may write

$$C_X = \text{closure of } (\Gamma_X \backslash \pi^{-1}(\operatorname{Sing} X)) \subset X \times \mathbb{P}_N^*.$$

Of course, it need not be locally trivial along the singular locus. In case of a hypersurface the space C_X is the graph of the rational map δ defined above. The result of the previous construction can be summarized in the following proposition.

Proposition. *Let $X \subset \mathbb{P}_N$ be a variety of dimension n, let $C_X \subset X \times \mathbb{P}_N^*$ be the conormal bundle defined above and denote by*

$$\pi^* : X \times \mathbb{P}_N^* \longrightarrow \mathbb{P}_N^*$$

the canonical projection. Define the dual varieties

$$X^* := \pi^*(C_X) \subset \mathbb{P}_N^* \quad and \quad X^\vee := \mathcal{D}(X^*) \subset \mathbb{P}_N^\vee.$$

Then:

1) C_X is irreducible and $\dim C_X = N - 1$,

2) $N - n - 1 \leq \dim X^* = \dim X^\vee \leq N - 1$,

3) $X^\vee = closure \ of \ \{H \in \mathbb{P}_N^\vee : H \supset \mathbb{T}_p X \ \ for \ some \ smooth \ p \in X\}.$

The lower bound for $\dim X^\vee$ in 2) is clear, since $X^\vee$ contains the linear space

$$\{H \in \mathbb{P}_N^\vee : \mathbb{T}_p X \subset H\}$$

of dimension $N - n - 1$ for every smooth $p \in X$. The lower bound is attained for a linear variety X, but "in general", the upper bound is attained, i.e. $X^\vee$ is a hypersurface [Kl]. In any case, the difference

$$\delta_*(X) := N - 1 - \dim X^*$$

is called the *dual defect* of X. Assertion *2)* implies

$$0 \leq \delta_*(X) \leq \dim X .$$

A variety X is called *ordinary*, if $\delta_*(X) = 0$. In 2.5, we will give more detailed results on the dual defect.

Example. If $X \subset \mathbb{P}_N$ is a smooth quadric, then there is an invertible, symmetric matrix $A \in \mathrm{GL}\,(n + 1, \mathbb{C})$ such that

$$X = \{x \in \mathbb{P}_N : {}^t x A x = 0\} .$$

Since the gradient of this equation is $2\,{}^t x A$, we have

$$\mathbb{T}_p X = \{x \in \mathbb{P}_N : {}^t p A x = 0\} \quad \text{for every } p \in X .$$

The matrix A can be used to give an isomorphism

$$\alpha : \mathbb{P}_N \longrightarrow \mathbb{P}_N^* , \quad x \longmapsto {}^t x A = y ,$$

and for the dual hypersurface we have $X^* = \alpha(X)$. This implies that

$$X^* = \left\{y \in \mathbb{P}_N^* : y\left(A^{-1}\right) {}^t y = 0\right\} .$$

2.1.5 Duality Theorem

The process of dualizing can be iterated: if $X \subset \mathbb{P}_N$, then $X^* \subset \mathbb{P}_N^*$ and $(X^*)^* \subset (\mathbb{P}_N^*)^* = \mathbb{P}_N$. We want to show that

$$(X^*)^* = X .$$

We prove slightly more (see [H, Kl]):

Duality Theorem. *If $X \subset \mathbb{P}_N$ is a variety and $X^* \subset \mathbb{P}_N^*$ is the dual variety, then we have conormal bundles*

$$C_X \subset X \times \mathbb{P}_N^* \subset \mathbb{P}_N \times \mathbb{P}_N^* \quad \text{and} \quad C_{X^*} \subset X^* \times \mathbb{P}_N \subset \mathbb{P}_N^* \times \mathbb{P}_N.$$

Then, up to the order of factors, $C_X = C_{X^*}$.

Proof. We show that an open subset of $\widehat{C}_X \subset \widehat{\Gamma}_X$ is also contained in $\widehat{C}_{X^*} \subset \widehat{\Gamma}_{X^*}$. This implies $C_X \subset C_{X^*}$ and by symmetry $C_{X^*} \subset C_X$. For this purpose, we consider the affine cones $\widehat{X} \subset \mathbb{C}^{N+1}$, $\widehat{X}^* \subset (\mathbb{C}^{N+1})^*$ and the algebraic sets

$$\widehat{\Gamma}_X \subset \widehat{X} \times \left(\mathbb{C}^{N+1}\right)^* \quad \text{with} \quad \widehat{\Gamma}_{X,x} = \left\{y \in \left(\mathbb{C}^{N+1}\right)^* : \langle T_x\widehat{X}, y\rangle = 0\right\} ,$$

$$\widehat{\Gamma}_{X^*} \subset \mathbb{C}^{N+1} \times \widehat{X}^* \quad \text{with} \quad \widehat{\Gamma}_{X^*,y} = \left\{x \in \left(\mathbb{C}^{N+1}\right) : \langle x, T_y\widehat{X}^*\rangle = 0\right\}$$

for smooth points $x \in \widehat{X}$ and $y \in \widehat{X}^*$ (2.1.4).

If $\xi : \widehat{C}_X \to (\mathbb{C}^{N+1})^*$ denotes the projection, then we take a smooth point $(p, q) \in \widehat{C}_X \subset \widehat{\Gamma}_X$ such that $p \in \widehat{X}$ and $q \in \widehat{X}^* = \xi(\widehat{C}_X)$ are also smooth, and the differential of ξ at (p, q) has rank dim $\widehat{X}^*$. In order to show $(p, q) \in \widehat{C}_{X^*} \subset \widehat{\Gamma}_{X^*}$, take an open set $0 \in U \subset \mathbb{C}^N$ and a local parameterization

$$\varphi : U \longrightarrow \widehat{C}_X \subset \widehat{\Gamma}_X, \quad (t_1, \ldots, t_N) \longmapsto (x(t), y(t)), \quad \varphi(0) = (p, q).$$

For $x \in \widehat{X}$ we know $x \in T_x\widehat{X}$; hence, the condition for $\widehat{\Gamma}_X$ implies $\langle x(t), y(t) \rangle = 0$ for every $t \in U$. Taking partial derivatives of this relation, we obtain

$$\left\langle \frac{\partial x}{\partial t_i}, y \right\rangle + \left\langle x, \frac{\partial y}{\partial t_i} \right\rangle = 0 \quad \text{for } i = 1, \ldots, N.$$

Now $\frac{\partial x}{\partial t_i} \in T_x\widehat{X}$; hence, the first bracket vanishes, and consequently the second bracket as well. In order to see that the condition for $\widehat{\Gamma}_{X^*}$ is satisfied, we have to prove that

$$\frac{\partial y}{\partial t_1}(0), \ldots, \frac{\partial y}{\partial t_N}(0) \quad \text{generate } T_q\widehat{X}^* \subset (\mathbb{C}^{N+1})^*.$$

This is an immediate consequence of the assumption on the differential of the projection ξ. Furthermore, the smoothness of $\widehat{X}^*$ in the point q implies that (p, q) lies not only in $\widehat{\Gamma}_{X^*}$, but in $\widehat{C}_{X^*}$. $\square$

Corollary. *Let $C_X \subset \mathbb{P}_N \times \mathbb{P}_N^*$ be the conormal bundle of a variety $X \subset \mathbb{P}_N$ and denote by*

$$\pi : C_X \longrightarrow \mathbb{P}_N, \quad \pi^* : C_X \longrightarrow \mathbb{P}_N^*$$

the canonical projections. Then

$$X^* = \pi^*(C_X), \quad (X^*)^* = \pi(C_X) = X.$$

2.1.6 The Contact Locus

As another immediate consequence of the Duality Theorem, we prove the *linearity of the contact locus*. If $X \subset \mathbb{P}_N$ is a variety with smooth locus $X_{\text{sm}} \subset X$ and $H \subset \mathbb{P}_N$ is a hyperplane then we define the *contact locus* [Kl]

$$X_H := \text{closure of } \{x \in X_{\text{sm}} : \mathbb{T}_x X \subset H\} \subset X.$$

To simplify the notation, we identify $\mathbb{P}_N^\vee$ and $\mathbb{P}_N^*$, i.e. a hyperplane $H \in \mathbb{P}_N^\vee$ and its linear equations $y \in \mathbb{P}_N^*$. Consequently,

$$\mathbb{T}_x X \subset H \iff \langle \mathbb{T}_x X, y \rangle = 0.$$

To dualize this condition, we use the conormal bundle and its projections

$$X \xleftarrow{\pi} C_X \xrightarrow{\pi^*} X^*.$$

If $y \notin X^*$, then $X_H = \emptyset$. Otherwise, we have

$$X_H = \pi \left(\pi^{*-1}(y) \right) = \text{closure of } \{ x \in X_{\text{sm}} : \langle \mathbb{T}_x X, y \rangle = 0 \} \neq \emptyset.$$

By 2.1.5, we know that, for $x \in X_{\text{sm}}$ and $y \in X^*_{\text{sm}}$,

$$\langle \mathbb{T}_x X, y \rangle = 0 \iff \langle x, \mathbb{T}_y X^* \rangle = 0 ;$$

hence,

$$X_H = \left\{ x \in \mathbb{P}_N : \langle x, \mathbb{T}_y X^* \rangle = 0 \right\} \subset X .$$

As a corollary, we obtain

BERTINI's **Theorem.** *If $X \subset \mathbb{P}_N$ is irreducible and $H \in X^\vee$ is a smooth point, then the contact locus $X_H \subset X$ is linear of dimension $N - 1 - \dim X^\vee = \delta_*(X)$.*

In general, the dual defect $\delta_*(X)$ is zero; in the above case the generic contact locus is a single point.

2.1.7 The Dual Curve

We consider the special case where $X \subset \mathbb{P}_N$ is a *curve*, i.e. a one-dimensional variety.

Remark. *A curve is ordinary iff it is not a line.*

Proof. If X is a line, then X^* is linear of codimension two. If X is not a line, then X^* contains infinitely many different two-codimensional linear spaces; hence, codim $X^* \leq 1$; so X is ordinary.   □

If the curve $X \subset \mathbb{P}_N$ is not only ordinary, but moreover *nondegenerate* (i.e. not contained in a proper linear subspace of $\mathbb{P}_N$), then we are going to construct a curve $\tilde{X} \subset \mathbb{P}_N^*$ derived from X which is in close relation to the hypersurface X^*.

Recall from 1.5.4 that a nondegenerate curve, which we consider in the parameterized form

$$\varphi : S \longrightarrow X \subset \mathbb{P}_N ,$$

has associated curves

$$\varphi^{(k)} : S \longrightarrow X^{(k)} \subset \mathbb{G}(k, N) ,$$

for $0 \leq k \leq N - 1$. The extreme case is $k = N - 1$; here we have

$$S \overset{\varphi^{(N-1)}}{\longrightarrow} \mathbb{G}(N - 1, N) \overset{\mathcal{D}}{\longrightarrow} \mathbb{G}^*(0, N) = \mathbb{P}_N^* ,$$

where $\mathcal{D} = \mathcal{D}_{N-1,N}$ is the isomorphism given by duality (1.3.4), and we call

$$\tilde{\varphi} := \mathcal{D} \circ \varphi^{(N-1)} : S \to \tilde{X} \subset \mathbb{P}_N^*$$

the *dual curve* of the curve φ, where the image $\tilde{X}$ contains all osculating hyperplanes of X (if we identify points of $\mathbb{P}_N^*$ with hyperplanes).

Now we can compare the associated curves of φ and $\tilde{\varphi}$ via duality, i.e. consider the diagram

$$
\begin{array}{ccc}
 & S & \\
\varphi^{(N-k-1)} \nearrow & & \searrow \tilde{\varphi}^{(k)} \\
\mathbb{G}(N-k-1, N) & \xrightarrow{\ D\ } & \mathbb{G}^*(k, N).
\end{array}
$$

For $k = 0$, it commutes; this is the definition of $\tilde{\varphi}$. By exchanging derivatives between φ and $\tilde{\varphi}$, we prove the

Theorem. *For a nondegenerate curve $\varphi : S \to \mathbb{P}_N$ and $0 \le k \le N - 1$ we have*

$$
D \circ \varphi^{(N-k-1)} = \tilde{\varphi}^{(k)}.
$$

Proof. We use induction and Proposition 1.5.2. For $k = 0$, this is the definition of $\tilde{\varphi}$. Assume the equality is true for $k - 1$, then

$$
\begin{aligned}
\tilde{\varphi}^{(k)} &= \left(\tilde{\varphi}^{(k-1)}\right)^{(1)} = \left(D \circ \varphi^{(N-(k-1)-1)}\right)^{(1)} \\
&= D \circ \left(\varphi^{(N-k)}\right)_{(1)} = D \circ \varphi^{(N-k-1)}.
\end{aligned}
$$
$\square$

If $k = N - 1$, then $D \circ \varphi = \tilde{\varphi}^{(N-1)}$. Dualizing, we get $\varphi = D \circ \tilde{\varphi}^{(N-1)} = \tilde{\tilde{\varphi}}$; hence, the following corollary

Corollary 1. $\tilde{\tilde{X}} = X$.

If $k = N - 2$, we have duality between $\varphi^{(1)}$ and $\tilde{\varphi}^{(N-2)}$; this means that the osculating $(N-2)$-plane $\tilde{\varphi}^{(N-2)}(s) \subset \mathbb{P}_N^*$ is the annihilator of the tangent line $\varphi^{(1)}(s)$ and that

$$
\bigcup_{s \in S} \tilde{\varphi}^{(N-2)}(s) = X^*
$$

is the hypersurface dual to X. But the union is, by definition, the osculating scroll of order $N - 2$ of $\tilde{X}$ (see 2.2.7, Example 3).

Corollary 2. $\mathrm{Tan}^{(N-2)}\tilde{X} = X^*$.

In particular, if $N = 3$, this implies that the dual surface $X^* \subset \mathbb{P}_3^*$ of $X \subset \mathbb{P}_3$ is the tangent scroll $\mathrm{Tan}\,\tilde{X}$ of the dual curve $\tilde{X} \subset \mathbb{P}_3^*$.

2.1.8 Rational Curves

For a nonlinear curve $X \subset \mathbb{P}_N$, the dual variety $X^* \subset \mathbb{P}_N^*$ is a hypersurface (2.1.6). For a nondegenerate rational curve, an equation of the dual hypersurface can be easily computed by a method which has already been used by CAYLEY [C].

Take any nondegenerate *rational curve* of degree $d \geq N$

$$\varphi : \mathbb{P}_1 \longrightarrow \mathbb{P}_N , \quad (t_0 : t_1) \longmapsto (\varphi_0(t_0, t_1) : \ldots : \varphi_N(t_0, t_1)) ,$$

where $\varphi_0, \ldots, \varphi_N$ are homogeneous polynomials of degree d without common zero. Important examples are the *rational monomial curves*: if $0 = m_0 < m_1 < \ldots < m_{N-1} < m_N = d$, then

$$\varphi_i(t_0, t_1) := t_0^{d - m_i} t_1^{m_i} .$$

Now, if $(y_0 : \ldots : y_N)$ are coordinates in $\mathbb{P}_N^*$, we consider the polynomial

$$\Phi(t; y) := \langle \varphi(t), y \rangle := y_0 \varphi_0(t) + \ldots + y_N \varphi_N(t) .$$

If we consider the y_i as coefficients and $(t_0 : t_1)$ as the homogeneous variable, there is a *discriminant set*

$$D := \left\{ (y_0 : \ldots : y_N) \in \mathbb{P}_N^* : \Phi(t; y) \text{ has a multiple zero } (t_0 : t_1) \in \mathbb{P}_1 \right\} .$$

D can be described explicitly as follows. Consider the sets

$$Z = \left\{ (t; y) \in \mathbb{P}_1 \times \mathbb{P}_N^* : \Phi(t; y) = 0 \right\} \quad \text{and}$$

$$Z' = \left\{ (t; y) \in \mathbb{P}_1 \times \mathbb{P}_N^* : \frac{\partial \Phi}{\partial t_0}(t; y) = \frac{\partial \Phi}{\partial t_1}(t; y) = 0 \right\} .$$

By EULER's formula $Z' \subset Z$. If we denote by

$$\pi : \mathbb{P}_1 \times \mathbb{P}_N^* \longrightarrow \mathbb{P}_N^*$$

the canonical projection, then $D = \pi(Z')$. We claim that $D = X^*$.

Consider the product map

$$\varphi \times \text{id} : \mathbb{P}_1 \times \mathbb{P}_N^* \longrightarrow \mathbb{P}_N \times \mathbb{P}_N^* .$$

By definition of X^* as projection of the conormal variety $C_X \subset \mathbb{P}_N \times \mathbb{P}_N^*$, it suffices to prove

$$(\varphi \times \text{id})(Z') = C_X ;$$

this follows immediately from

$$\langle \mathbb{T}_{\varphi(t)} X, y \rangle = 0 \iff \left\langle \frac{\partial \varphi}{\partial t_0}(t), y \right\rangle = \left\langle \frac{\partial \varphi}{\partial t_1}(t), y \right\rangle = 0 \iff \frac{\partial \Phi}{\partial t_0}(t; y) = \frac{\partial \Phi}{\partial t_1}(t; y) = 0 .$$

So, finally, an equation for X^* can be obtained as a discriminant.

Example. Take the twisted cubic

$$\varphi : \mathbb{P}_1 \longrightarrow X \subset \mathbb{P}_3, \quad (t_0 : t_1) \mapsto \left(t_0^3 : t_0^2 t_1 : t_0 t_1^2 : t_1^3 \right),$$

and consider the inhomogeneous polynomial

$$\varphi(t; y) := y_0 t^3 + y_1 t^2 + y_2 t + y_3 .$$

Its discriminant up, to a factor of y_0, is

$$27 \, y_0^2 \, y_3^2 - 18 \, y_0 \, y_1 \, y_2 \, y_3 + 4 \, y_0 \, y_2^3 + 4 \, y_1^3 \, y_3 - y_1^2 \, y_2^2 ,$$

and this is the equation for $X^* = \mathrm{Tan} \, \tilde{X} \subset \mathbb{P}_3^*$.

The dual curve of φ is

$$\tilde{\varphi} : \mathbb{P}_1 \longrightarrow \tilde{X} \subset \mathbb{P}_3^*, \quad (t_0 : t_1) \mapsto \left(t_0^3 : -3 \, t_0^2 t_1 : 3 \, t_0 t_1^2 : -t_1^3 \right),$$

because, by definition, $\tilde{\varphi} := \mathcal{D} \circ \varphi^{(2)}$ and

$$\varphi^{(2)}(t) \;=\; \mathbb{P} \left(\mathrm{span} \left\{ \frac{\partial^2 \varphi}{\partial t_0^2}, \frac{\partial^2 \varphi}{\partial t_0 \partial t_1}, \frac{\partial^2 \varphi}{\partial t_1^2} \right\} \right)$$

$$=\; \mathbb{P} \left(\mathrm{span} \left\{ \begin{pmatrix} 6 t_0 \\ 2 t_1 \\ 0 \\ 0 \end{pmatrix}, \begin{pmatrix} 0 \\ 2 t_0 \\ 2 t_1 \\ 0 \end{pmatrix}, \begin{pmatrix} 0 \\ 0 \\ 2 t_0 \\ 6 t_1 \end{pmatrix} \right\} \right).$$

By the same computation as above and changing only the coordinates by $x_0 = y_0$, $3 x_1 = -y_1$, $3 x_2 = y_2$, $x_3 = -y_3$, we get the equation

$$x_0^2 x_3^2 - 6 x_0 x_1 x_2 x_3 + 4 x_0 x_2^3 + 4 x_1^3 x_3 - 3 x_1^2 x_2^2$$

for the tangent scroll $\mathrm{Tan} \, X = \tilde{X}^* \subset \mathbb{P}_3$.

The reader may compute the corresponding equations for the possible monomial curves of degree 4 in $\mathbb{P}_3$ and compare the results with [C].

2.2 Developable Varieties

In this section, we introduce the basic concepts of rulings and developability of varieties.

2.2.1 Rulings

An easy method to construct varieties is to consider unions of families of linear spaces. In the terminology of [H], these varieties are "swept out" by linear spaces. The basic recipe is given in 2.1.1: For a variety $B \subset \mathbb{G}(k, N)$, we consider the incidence variety

$$I_B \subset \mathbb{P}_N \times \mathbb{G}(k, N) \quad \text{and its projection } \pi : I_B \longrightarrow \mathbb{P}_N .$$

Then we have seen that

$$X := \bigcup_{E \in B} E = \pi(I_B) \subset \mathbb{P}_N$$

is a variety, and $\dim X \leq \dim B + k$. This bound is called the *expected dimension*. We call B the *base* and every linear space $E \in B$ a *leaf* of the family.

Without further conditions, a general point of X may lie on too many linear spaces, so the first restriction we have to impose is evident:

we say X is *uniruled* by (k-planes of) B if it has the expected dimension, i.e.

$$\dim X = \dim B + k .$$

Since, in any case, $\dim I_B = \dim B + k$, this condition implies $\dim I_B = \dim X$; hence, π is generically finite, and consequently, there is a Zariski-open subset $U \subset X$ such that every $p \in U$ is contained in only finitely many leaves.

There are evident improvements, which can be expressed in terms of the projection $\pi : I_B \to X$:

- X is called *almost ruled* if π is birational.

- X is called *ruled* if π is biregular.

This means that almost every (resp. every) point of X is contained in a single leaf.

If we use the notation

$$B_p := \{E \in B : p \in E\} \subset B$$

for $p \in X$, then X is

uniruled	$\Longleftrightarrow$	B_p is finite for almost every $p \in X$,
almost ruled	$\Longleftrightarrow$	B_p is a single leaf for almost every $p \in X$,
ruled	$\Longleftrightarrow$	B_p is a single leaf for every $p \in X$.

Of course a variety may have several rulings (like a quadric in $\mathbb{P}_3$), but these conditions are very strong. A first elementary result in this direction is the

Proposition. *Let $X \subset \mathbb{P}_N$ be a variety ruled by k-planes with base B, and assume that* $\dim B > 0$. *Then*

$$\operatorname{codim} X \geq k \,.$$

Proof. We choose a point $L \in B$ and a hypersurface $Z \subset B$ with $L \notin Z$. Then, in $\mathbb{P}_N$, the varieties L and

$$Y := \bigcup_{E \in Z} E$$

are disjoint. The dimension formula for the intersection implies

$$N > \dim L + \dim Y = k + (k + n - k - 1) = k + n - 1 \,,$$

where $n := \dim X$; hence,

$$\operatorname{codim} X = N - n > k - 1 \,. \qquad \square$$

For uniruled X, the restriction on the codimension is not so strong (2.4.5); for example, a smooth quadric hypersurface $X \subset \mathbb{P}_N$ is uniruled by $\lfloor \frac{N-1}{2} \rfloor$-planes.

Example 1. Take $X = \mathbb{P}_2$ and

$$B = \{(y_0 : y_1 : y_2) \in \mathbb{P}_2^* : y_0 = 0\} \,.$$

Then we have $B_p = B$, for $p = (1 : 0 : 0)$ and B_q consists of a single "leaf" for $q \neq p$. $\mathbb{P}_2$ is almost ruled by B, and the variety I_B is the "blow up" of $\mathbb{P}_2$ in p.

For "generic" B, the set B_p contains a single linear subspace for $p \in U$, but this is not always the case.

Example 2. Take $X = \mathbb{P}_2$ and

$$B = \{(y_0 : y_1 : y_2) \in \mathbb{P}_2^* : y_1^2 = y_0\, y_2\} \subset \mathbb{P}_2^* \cong \mathbb{G}(1, 2) \,.$$

Then B_p consists of the tangents to the conic

$$\{(x_0 : x_1 : x_2) \in \mathbb{P}_2 : x_1^2 = 4\, x_0\, x_2\}$$

passing p, and $\mathbb{P}_2$ is not uniruled by B.

For any k, we consider

$$X_k := \bigcup_{E \subset \mathrm{F}_k(X)} E \subset X \,.$$

In general, $X_k \subset X$ is a proper subset, but if there is a ruling, we have $B \subset \mathrm{F}_k(X)$ and $X_k = X$. This shows that rulings exist only if the Fano variety is sufficiently large.

2.2.2 Adapted Parameterizations

In 2.2.1, the variety $X \subset \mathbb{P}_N$ swept out by $B \subset \mathbb{G}(k, N)$ was defined as the image of a regular projection map. For a more detailed study of X, we need good local descriptions.

Let us start with a smooth point $E \in B$. If $r = \dim B$, then there is an open neighborhood S of $o \in \mathbb{C}^r$ and a parameterization (i.e. a biholomorphic map)

$$\psi : S \longrightarrow V \subset B$$

of a neighborhood V of E in B. Now we can define the concepts of moving vectors and a local basis as we have done it in 1.5.1 for a curve B: a family $(\varrho_0, \dots, \varrho_k)$ of holomorphic maps,

$$\varrho_i : S \longrightarrow \mathbb{C}^{N+1},$$

is called a *local basis for B* (or *of the uniruling* if this is the case) if

$$(\varrho_0(s), \dots, \varrho_k(s)) \quad \text{is a basis of } \psi(s) \text{ for every } s \in S.$$

The existence can be proved exactly as in 1.5.1, and if ψ is given explicitly, it can be computed. By definition of X, the image of the holomorphic map

$$\widehat{\Phi} : S \times \mathbb{C}^{k+1} \longrightarrow \mathbb{C}^{N+1}, \quad (s, t_0, \dots, t_k) \longmapsto \sum_{i=0}^{k} t_i \varrho_i(s),$$

is contained in $\widehat{X}$, the affine cone of X. The Jacobian of $\widehat{\Phi}$ is

$$\left(\sum_{i=0}^{k} t_i \frac{\partial \varrho_i}{\partial s_1}, \dots, \sum_{i=0}^{k} t_i \frac{\partial \varrho_i}{\partial s_r}, \varrho_0, \dots, \varrho_k \right).$$

If there is a point $(s, t) \in S \times \mathbb{C}^{k+1}$ such that

$$\text{rank Jac}_{(s,t)} \widehat{\Phi} = r + k + 1,$$

then X has the expected dimension, and thus, B determines a uniruling.

In 2.2.1, we considered the incidence variety I_B with its projections

$$X \xleftarrow{\pi} I_B \xrightarrow{\xi} B.$$

The projective version of $\widehat{\Phi}$ is

$$\Phi : S \times \mathbb{P}_k \longrightarrow \pi(\xi^{-1}(V)) \subset X \subset \mathbb{P}_N, \quad (s; t_0 : \dots : t_k) \longmapsto \mathbb{C}\left(\sum_{i=0}^{k} t_i \varrho_i(s) \right).$$

For every $s \in S$, the restriction of Φ

$$\Phi_s : \{s\} \times \mathbb{P}_k \longrightarrow \psi(s) \subset X$$

is a linear parameterization of the leaf.

The quality of Φ in the direction of S depends on the properties of the projection π. If X is uniruled by B, then for almost all $(p, E) \in I_B$, there is neighborhood $W \subset I_B$ such that the restriction of π,

$$\pi' : W \longrightarrow U \subset X \, ,$$

is biholomorphic. Hence, the corresponding restriction of Φ is a parameterization of U, and we call Φ an *adapted parameterization* of the uniruling.

If X is almost ruled or even ruled by B, then W and U can be chosen correspondingly larger.

In an affine part, the parameterization looks as follows: if $p = (1 : p_1 : \dots : p_N) \in E$, then we can assume

$$\varrho_0(s) = (1, \varphi(s)) \text{ with } \varphi(o) = (p_1, \dots, p_N) \text{ and } \varrho_j(s) = (0, \sigma_j(s))$$

for $j = 1, \dots, N$, where $\varphi, \sigma_j : S \to \mathbb{C}^N$ are holomorphic maps. In this case, the restriction of Φ gives a map

$$\Phi' : S \times \mathbb{C}^k \longrightarrow X' := \{(x_0 : x_1 : \dots : x_N) \in X : x_0 \neq 0\} \subset \mathbb{C}^N \, ,$$

$$(s, t_1, \dots, t_k) \longmapsto \varphi(s) + \sum_{i=1}^{k} t_i \sigma_i(s) \, .$$

If the above projection π was biholomorphic in (p, E), then Φ' is biholomorphic in the point $(o, o) \in S \times \mathbb{C}^k$, i.e. the rank of the Jacobian of Φ' at $o = (o, o)$

$$\left(\frac{\partial \varphi}{\partial s_1}(o), \dots, \frac{\partial \varphi}{\partial s_r}(o), \sigma_1(o), \dots, \sigma_k(o) \right)$$

is equal to $r + k = \dim X$.

The subvariety $\Phi'(S \times \{o\}) = \varphi(S)$ is called a *local directrix* of the ruling.

2.2.3 Germs of Rulings

As we have seen, a ruling defined via a base in the Grassmannian gives a foliation in most points of the variety. We take a look at the inverse problem: given a holomorphic "germ of a ruling" in some open set, can it be extended to the entire space, i.e. is there a base in the Grassmannian? In general, such an extension is impossible since the corresponding germ in the Fano variety need not be algebraic; however, this is true if the dimension of the Fano variety is sufficiently small.

Proposition. *Let $X \subset \mathbb{P}_N$ be a variety of dimension n. Assume there is an open and smooth U in some affine part $X' \subset \mathbb{C}^N$ with a parameterization*

$$\Phi : S \times T \longrightarrow U \subset X' \, ,$$

$$(s, t_1, \dots, t_k) \longmapsto \varphi(s) + \sum_{i=1}^{k} t_i \sigma_i(s) \, ,$$

where $S \subset \mathbb{C}^r$ and $T \subset \mathbb{C}^k$ are open neighborhoods of the origin, $k + r = n$, and

$$\varphi, \sigma_1, \ldots, \sigma_k : S \longrightarrow \mathbb{C}^N$$

are holomorphic maps. If $\dim F_k(X) \leq r$, then Φ induces a unique k-dimensional uniruling of X.

Proof. For every $s \in S$, we have $\Phi(\{s\} \times T) \subset U$; hence, the projective closure $E_s \subset \mathbb{P}_N$ is contained in X. Thus, Φ determines an injective holomorphic map

$$\psi : S \longrightarrow F_k(X).$$

By standard arguments, this implies $\dim F_k(X) \geq r$, and our assumption yields $\dim F_k(X) = r$. Hence, there is a unique irreducible component $B \subset F_k(X)$ with $\dim B = r$ and $\psi(S) \subset B$. This is the base of the uniruling. $\qquad\qquad\square$

2.2.4 Developable Rulings and Focal Points

Let $X \subset \mathbb{P}_N$ be a variety of dimension n with a k-dimensional uniruling given by its base $B \subset \mathbb{G}(k, N)$ of dimension r; by definition (2.2.1), we have

$$n = k + r.$$

The uniruling is called *developable* if the tangent space $\mathbb{T}_p X \subset \mathbb{P}_N$ is the same for all smooth points $p \in X$ lying in the same leaf $E \in B$.

In order to obtain a simple criterion for developability in terms of the rank of a matrix, we use the concept of a local basis of the ruling for a local parameterization

$$\psi : S \longrightarrow B,$$

with $S \subset \mathbb{C}^r$ open (2.2.2). If $(\varrho_0, \ldots, \varrho_k)$ is such a basis near a smooth point of B, we consider the $(k + 1)(r + 1)$ holomorphic maps

$$\varrho_i, \frac{\partial \varrho_i}{\partial s_j} : S \longrightarrow \mathbb{C}^{N+1}, \quad i = 0, \ldots, k, \quad j = 1, \ldots, r,$$

and the span of their values. Now we can see how restrictive the condition of developability is.

Developability Criterion. *If the uniruling $B \subset \mathbb{G}(k, N)$ of $X \subset \mathbb{P}_N$ is developable, then for every local basis $(\varrho_0, \ldots, \varrho_k)$ near a smooth point of B we have*

$$\operatorname{rank} \left(\varrho_i, \frac{\partial \varrho_i}{\partial s_j} \right)_{i,j} \leq \dim X + 1 \qquad\qquad (*)$$

everywhere on S.

If, conversely, there is one local basis near every smooth point of B satisfying this rank condition, then the uniruling is developable.

In particular, every other local basis satisfies the same rank condition.

The simplest case is $r = 1$. If we write $\varrho_i' = \frac{d\varrho_i}{ds}$, then the developability is equivalent to

$$\text{rank}\,(\varrho_0, \dots, \varrho_k, \varrho_0', \dots, \varrho_k') = k + 2$$

almost everywhere on S. In the terminology of 1.5.1, this means that the curve in $\mathbb{G}(k, N)$ has drill one.

Proof. A local basis $(\varrho_0, \dots, \varrho_k)$ yields a local parameterization

$$\Phi : S \times \mathbb{C}^{k+1} \longrightarrow \widehat{X} \subset \mathbb{C}^{N+1}, \quad (s, t_0, \dots, t_k) \mapsto \sum_{i=0}^{k} t_i \varrho_i(s),$$

of the affine cone of X. Therefore, the tangent space of $\widehat{X}$ in smooth points is spanned by the columns of the Jacobian of Φ. They are

$$\sum_{i=0}^{k} t_i \frac{\partial \varrho_i}{\partial s_1}, \dots, \sum_{i=0}^{k} t_i \frac{\partial \varrho_i}{\partial s_r}, \varrho_0, \dots, \varrho_k.$$

Further, consider the distinguished points

$$q_i(s) := \Phi(s, 0, \dots, 1, \dots, 0),$$

where 1 is in the position of t_i. If $\widehat{X}$ is smooth in $q_i(s)$, then

$$T_{q_i(s)}\widehat{X} = \text{span} \left\{ \varrho_0(s), \dots, \varrho_k(s), \frac{\partial \varrho_i}{\partial s_1}(s), \dots, \frac{\partial \varrho_i}{\partial s_r}(s) \right\}.$$

If we assume X is developable, then for every s the tangent spaces $T_{q_i(s)}\widehat{X}$ have to be the same for $i = 0, \dots, k$, provided $\widehat{X}$ is smooth in all these points. This immediately implies the rank condition $(*)$.

Conversely, take a smooth point $E \in B$ such that X has smooth points on E, consider a local basis $(\varrho_0, \dots, \varrho_k)$ near $E = \psi(o)$, and define

$$T_E := \text{span} \left\{ \varrho_i(o), \frac{\partial \varrho_i}{\partial s_j}(o) \right\}_{i,j} \subset \mathbb{C}^{N+1}.$$

Condition $(*)$ implies $\dim T_E \le k + r + 1$. Now, if $p \in E \subset X$ is a smooth point, and $q \in \widehat{X}$ is lying over p, then $q = \Phi(o, t)$ for some $t \in \mathbb{C}^{k+1}$. As we have seen above, the tangent space $T_q\widehat{X}$ is spanned by the columns of the Jacobian; hence, for any such q we have $T_q\widehat{X} \subset T_E$. In fact, the equality $T_q\widehat{X} = T_E$ holds, since $\dim T_q\widehat{X} = \dim \widehat{X} = k+r+1$. □

It is obvious that developability is an extremely strong condition for a uniruling. More generally, one can measure the variation of the tangent space along a leaf by considering the semicontinuous function

$$\dim \mathrm{span} \left\{ \varrho_i, \frac{\partial \varrho_i}{\partial s_j} \right\}_{i,j} - (k+1)$$

and introducing a "drill" of the ruling B; this has been done in 1.5.1 in the case of a curve B. It is fairly clear that such a drill is an invariant of B, which is independent of the local basis. Developability is just the extreme case of drill $\dim B$. This also shows that developability can already be discovered on an arbitrarily small open set of X.

The name *parameterization* for the above map

$$\Phi : S \times \mathbb{C}^{k+1} \longrightarrow \widehat{X} \subset \mathbb{C}^{N+1}$$

is justified only locally. In general, Φ need not be an immersion everywhere, and in case the uniruling is developable, this is impossible. This is the general phenomenon that developability forces singularities. We use the notation

$$\mathrm{Sing}\, \Phi := \{(s, t) \in S \times \mathbb{C}^{k+1} : \Phi \text{ not immersive at } (s, t)\} \,.$$

Proposition. *Let $X \subset \mathbb{P}_N$ be a variety with a developable uniruling given by its base $B \subset \mathbb{G}(k, N)$ of dimension r. Take a local parameterization*

$$\psi : S \longrightarrow B$$

with basis $(\varrho_0, \ldots, \varrho_k)$ and the corresponding holomorphic map

$$\Phi : S \times \mathbb{C}^{k+1} \longrightarrow \widehat{X} \subset \mathbb{C}^{N+1}, \quad (s, t) \longmapsto \sum t_i \varrho_i(s) \,.$$

Then there is a $(r \times r)$-matrix $A(s, t)$, holomorphic in s and linear in t, such that

$$\mathrm{Sing}\, \Phi = \{(s, t) \in S \times \mathbb{C}^{k+1} : \det A(s, t) = 0\} \,.$$

In particular, for every $s \in S$,

$$\mathrm{Sing}\, \Phi \cap \left(\{s\} \times \mathbb{C}^{k+1} \right)$$

is a hypersurfaces of degree r.

Proof. We may assume that Φ is immersive in $(s, e_0) := (s, 1, 0, \ldots, 0)$ for every $s \in S$. Hence, the developability implies

$$\mathrm{span} \left\{ \varrho_0, \ldots, \varrho_k, \sum_i t_i \frac{\partial \varrho_i}{\partial s_1}, \ldots, \sum_i t_i \frac{\partial \varrho_i}{\partial s_r} \right\} \subset \mathrm{span} \left\{ \varrho_0, \ldots, \varrho_k, \frac{\partial \varrho_0}{\partial s_1}, \ldots, \frac{\partial \varrho_0}{\partial s_r} \right\} \,.$$

The second set of vectors is a basis of $T_s := T_{\Phi(s, e_0)}\widehat{X}$ for every s, and we have equality iff Φ is immersive in (s, t). If we denote

$$\frac{\partial \varrho_i}{\partial s} := \left(\frac{\partial \varrho_i}{\partial s_1}, \ldots, \frac{\partial \varrho_i}{\partial s_r}\right) \quad \text{for } i = 0, \ldots, k,$$

then there are $(r \times r)$-matrices A_i, holomorphic in s, such that

$$\frac{\partial \varrho_i}{\partial s} = \frac{\partial \varrho_0}{\partial s} \cdot A_i \quad \text{mod } \psi(s).$$

If we define

$$A(s, t) := \sum_{i=0}^{k} t_i A_i(s),$$

then this matrix defines an endomorphism of $T_s/\psi(s) \cong \text{span}\left\{\frac{\partial \varrho_0}{\partial s_1}(s), \ldots, \frac{\partial \varrho_0}{\partial s_r}(s)\right\}$ by

$$\frac{\partial \varrho_0}{\partial s}(s) \longmapsto \frac{\partial \varrho_0}{\partial s}(s) \cdot A(s, t)$$

for every (s, t). Hence, $(s, t) \in \text{Sing } \Phi$ is equivalent to

$$\text{rank } A(s, t) < k \iff \det A(s, t) = 0. \qquad \square$$

It is easy to see that the above endomorphism of $T_s/\psi(s)$ can be described in an invariant way, i.e. without the use of a local basis. This implies that the hypersurface

$$\text{Foc}(X, \psi(s)) := \mathbb{P}\left(\Phi\left(\text{Sing } \Phi \cap \left(\{s\} \times \mathbb{C}^{k+1}\right)\right)\right) \subset \psi(s) \subset X$$

is independent of the choices made. It is called the *focal hypersurface* in the leaf $\psi(s)$. The algebraic set

$$\text{Foc } X := \text{closure of } \bigcup_{s \in B} \text{Foc}(X, \psi(s)) \subset X$$

is called the *focal variety* of X. It can be shown [L$_1$, 5.2] that

$$\text{Foc } X \subset \text{Sing } X.$$

In the classical case of developable surfaces in the three-space, the focal variety is just the critical directrix (0.4,0.5). Of course, not every singular point of X must be focal. The main source of additional singularities are intersections of leaves.

Obviously, we have $\dim \text{Foc}(X, \psi(s)) = k - 1$; hence,

$$k - 1 \leq \dim \text{Foc } X \leq k + r - 1 = \dim X - 1.$$

2.2.5 Developability of Joins

Our first series of examples is concerned with joins of various kinds.

Example 1. A *linear subspace* $X \subset \mathbb{P}_N$ has k-dimensional almost rulings for $0 \le k \le \dim X$ and many of them for $0 < k < \dim X$. They are all developable since $\mathbb{T}_p X = X$ for every $p \in X$. This trivial case will be excluded in most of the subsequent considerations.

Example 2. Next we take a variety Y of dimension $n-1$ and a point $p \in \mathbb{P}_N \backslash Y$. According to (2.1.3), the join variety

$$X := Y \# p$$

is the *cone over Y* with *vertex p*. The map

$$J : Y \longrightarrow \mathbb{G}(1, N), \quad y \longmapsto \overline{yp},$$

is regular since $p \notin Y$, and the image $B := J(Y) \subset \mathbb{G}(1, N)$ is the base of the canonical 1-dimensional almost ruling of the cone X. The developability of this almost ruling is an almost immediate consequence of the following special case of the TERRACINI Lemma:

Lemma. *If $X = Y \# p \subset \mathbb{P}_N$ is a cone as above, $y \in Y$ is smooth, and $x \in \overline{yp}$ is smooth in X, then*

$$\mathbb{T}_x X = \mathbb{T}_y Y \# p.$$

Proof. We use the affine cones $\widehat{Y} \subset \widehat{X} \subset \mathbb{C}^{N+1}$. If $\tilde{y} \in \widehat{Y} \backslash \{0\}$ is a smooth point over q, then we can find a domain $o \in S \subset \mathbb{C}^r$ with coordinates $s_1, \dots, s_r$ and a holomorphic map

$$\varrho : S \longrightarrow \widehat{Y} \subset \mathbb{C}^{N+1} \quad \text{with } \varrho(o) = \tilde{y},$$

such that the map

$$\varphi : S \times \mathbb{C} \longrightarrow \widehat{Y}, \quad (s, t_0) \longmapsto t_0 \varrho(s),$$

is a parameterization of $\widehat{Y}$ around $\tilde{y}$. If we take a point $\tilde{p} \neq 0$ over p, then the map

$$\psi : S \times \mathbb{C}^2 \longrightarrow \widehat{X}, \quad (s, t_0, t_1) \mapsto t_0 \varrho(s) + t_1 \tilde{p},$$

is a local parameterization for $t_0 \neq 0$. Consequently, the tangent vector spaces of $\widehat{Y}$ in $\tilde{y} = \varphi(o, 1)$ and $\widehat{X}$ in $\tilde{x} = \psi(o, 1, t_1)$ are given by

$$T_{\tilde{y}} \widehat{Y} = \mathrm{span} \left\{ \frac{\partial \varrho}{\partial s_1}(o), \dots, \frac{\partial \varrho}{\partial s_r}(o), \varrho(o) \right\} \quad \text{and}$$

$$T_{\tilde{x}} \widehat{X} = \mathrm{span} \left\{ \frac{\partial \varrho}{\partial s_1}(o), \dots, \frac{\partial \varrho}{\partial s_r}(o), \varrho(o), \tilde{p} \right\}.$$

Passing to $\mathbb{P}_N$, this implies the assertion. □

An immediately consequence is the

Corollary. *If the cone $X = Y \# p \subset \mathbb{P}_N$ is not a linear space then it is singular in p.*

Proof. By continuity, we can conclude from the Lemma that

$$\mathbb{T}_z X \supset \mathbb{T}_y Y \# p = \mathbb{T}_x X$$

for any point z along the line $\overline{yp}$. In particular, $\mathbb{T}_p X \supset \mathbb{T}_x X$ for points x of the open set $U = X_{\mathrm{sm}} \cap (Y_{\mathrm{sm}} \# p)$. We consider the following two cases.

First, if $\mathbb{T}_p X = \mathbb{T}_x X$ for all these x on an open subset U, then $\mathbb{T}_p X = \mathbb{T}_x X$ for all $x \in X$, and X is a linear subspace.

Second, if there exist two different $\mathbb{T}_x X$ for $x \in U$ then the tangent space $\mathbb{T}_p X$ which contains these must have a bigger dimension than that of X; hence, X is singular in p. $\square$

Example 3. We take the cone $X = Y \# p$ as in the preceding example, but in addition we assume that Y has a developable k-dimensional uniruling with $0 \le k \le \dim Y$. If $E \subset Y$ is a leaf of the uniruling, then the tangent spaces $\mathbb{T}_y Y$ are the same for all $y \in E$. By the previous lemma, we know

$$\mathbb{T}_x X = \mathbb{T}_y Y \# p,$$

for smooth points $y \in Y$ and $x \in X$, so $\mathbb{T}_x X$ is the same for all smooth points $x \in E \# p$. This proves the following proposition.

Proposition. *A k-dimensional developable uniruling of a variety $Y \subset \mathbb{P}_N$ determines a developable $(k + 1)$-dimensional uniruling of the cone $X = Y \# p$ if $p \notin Y$.*

Now it is clear how this construction can be iterated: if $E \subset \mathbb{P}_N$ is a linear subspace of dimension e, disjoint from Y, then there are points $p_0, \ldots, p_e \in E$ such that

$$E = p_0 \# p_1 \# \ldots \# p_e \quad \text{and} \quad Y \# E = Y \# p_0 \# \ldots \# p_e.$$

$X = Y \# E$ is called the *cone over Y with vertex E*. If $y \in Y$ and $x \in y \# E$ are smooth points, then

$$\mathbb{T}_x X = \mathbb{T}_y Y \# E.$$

Hence, we obtain the following corollary.

Corollary. *Take a variety $Y \subset \mathbb{P}_N$ and a disjoint linear space $E \subset \mathbb{P}_N$. Then a developable k-dimensional uniruling of Y determines a developable uniruling of dimension $k + \dim E + 1$ of the cone $Y \# E$.*

Example 4. Cones are joins of a variety with a linear space. Now we consider the general case

$$X \# Y \subset \mathbb{P}_N \quad \text{for } X, Y \subset \mathbb{P}_N,$$

where X and Y are varieties with $\dim X = m$ and $\dim Y = n$. Let $\widehat{X}, \widehat{Y} \subset \mathbb{C}^{N+1}$ be their affine cones, $\widehat{p} \in \widehat{X}$ and $\widehat{q} \in \widehat{Y}$ smooth points with $\mathbb{C}\widehat{p} \ne \mathbb{C}\widehat{q}$. We use local parameterizations of a special form as follows. Take open neighborhoods of the origin

$$o \in S \subset \mathbb{C}^m, \quad s = (s_1, \ldots, s_m) \in S \quad \text{and} \quad o \in T \subset \mathbb{C}^n, \quad t = (t_1, \ldots, t_n) \in T,$$

holomorphic maps

$$\varrho : S \longrightarrow \widehat{X} \quad \text{and} \quad \sigma : T \longrightarrow \widehat{Y}$$

such that the maps

$$\varphi : S \times \mathbb{C} \longrightarrow \widehat{X}, \quad (s, \lambda) \longmapsto \lambda \varrho(s), \quad \psi : T \times \mathbb{C} \longrightarrow \widehat{Y}, \quad (t, \mu) \longmapsto \mu \sigma(t),$$

are parameterizations around $\widehat{p}$ and $\widehat{q}$. They yield a surjective map

$$\chi : S \times T \times \mathbb{C}^2 \longrightarrow (X \# Y)^{\wedge}, \quad (s, t, \lambda, \mu) \mapsto \lambda \varrho(s) + \mu \sigma(t).$$

For the tangent vector spaces in general points

$$\tilde{x} = \varphi(s, \lambda) \in \widehat{X}, \quad \tilde{y} = \psi(t, \mu) \in \widehat{Y}, \quad \text{and} \quad \tilde{z} = \chi(s, t, \lambda, \mu) \in (X \# Y)^{\wedge},$$

this implies

$$T_{\tilde{x}}\widehat{X} = \text{span} \left\{ \varrho(s), \frac{\partial \varrho}{\partial s_i}(s) : i = 1, \ldots, m \right\},$$

$$T_{\tilde{y}}\widehat{Y} = \text{span} \left\{ \sigma(t), \frac{\partial \sigma}{\partial t_j}(t) : j = 1, \ldots, n \right\}, \quad \text{and}$$

$$T_{\tilde{z}}(X \# Y)^{\wedge} = \text{span} \left\{ \varrho(s), \sigma(t), \frac{\partial \varrho}{\partial s_i}(s), \frac{\partial \sigma}{\partial t_j}(t) \right\}; \quad \text{hence},$$

$$T_{\tilde{z}}(X \# Y)^{\wedge} = T_{\tilde{x}}\widehat{X} + T_{\tilde{y}}\widehat{Y}.$$

Passing down to $\mathbb{P}_N$, this implies the [FOV]

Lemma of TERRACINI. *If $X, Y \subset \mathbb{P}_N$ are varieties, $x \in X$, $y \in Y$, $x \neq y$, and $z \in \overline{xy} \subset X \# Y$ are general points, then*

$$\mathbb{T}_z(X \# Y) = \mathbb{T}_x X \# \mathbb{T}_y Y.$$

As an immediate consequence, the canonical one-dimensional uniruling of $X \# Y$ is developable.

2.2.6 Dual Varieties of Cones and Degenerate Varieties

In the previous examples, we had defined cones in an explicit way as the join of an arbitrary variety with a disjoint linear subspace. We may also call a given space $X \subset \mathbb{P}_N$ a *cone* if there is a nonempty linear subspace $E \subset X$ such that for every $x \in X$ and $p \in E$ the line $\overline{xp}$ is contained in X. We want to show that there are many choices of subvarieties $Y \subset X$ such that $X = Y \# E$, and that there is a unique maximal *vertex* E.

The first part is straightforward: just take a linear subspace $L \subset \mathbb{P}_N$ such that $L \cap E = \emptyset$ and

$$\dim L + \dim E = N - 1,$$

then $Y := X \cap L$ has the property $X = Y \# E$. The second part can be seen by using the dual variety $X^* \subset \mathbb{P}_N^*$ (2.1.4). Recall that for a linear subspace $E \subset \mathbb{P}_N$ the dual $E^* \subset \mathbb{P}_N^*$ is also linear and

$$\dim E + \dim E^* = N - 1.$$

Also note that $Y \subset X$ does not necessarily imply $X^* \subset Y^*$. For simplicity, we consider points in $\mathbb{P}_N^*$ as hyperplanes in $\mathbb{P}_N$.

Proposition. *For a variety $X \subset \mathbb{P}_N$ and a linear subspace $E \subset \mathbb{P}_N$ the following conditions are equivalent:*

  i) X is a cone with vertex $E \subset X$,

  ii) $E \subset \mathbb{T}_x X$ for all smooth points $x \in X$,

  iii) $E \subset H$ for all $H \in X^*$,

  iv) $X^* \subset E^*$.

Proof. The implications *i) $\Rightarrow$ ii) $\Rightarrow$ iii) $\Rightarrow$ iv)* are clear. To prove *iv) $\Rightarrow$ i)* we use duality. A point $p \in E$ can be considered as a hyperplane in $\mathbb{P}_N^*$ such that

$$E^* \subset p \subset \mathbb{P}_N^*, \qquad\qquad (p)$$

and for almost every $x \in X$ we can find a smooth point $y \in X^*$ such that

$$\mathbb{T}_y X^* \subset x \subset \mathbb{P}_N^*. \qquad\qquad (x)$$

Now $X^* \subset E^*$ implies $\mathbb{T}_y X^* \subset E^*$, so by (x) and (p)

$$\mathbb{T}_y X^* \subset x \cap p \subset \mathbb{P}_N^*; \quad \text{hence,} \quad \overline{xp} \subset X^{**} = X \subset \mathbb{P}_N. \qquad \square$$

Corollary. *A variety $X \subset \mathbb{P}_N$ is a cone if and only if the dual variety $X^* \subset \mathbb{P}_N^*$ is contained in a proper linear subspace.*

The dual picture is the dual variety of a *degenerate* variety, i.e. a variety $Y \subset L \subset \mathbb{P}_N$, where L is a proper linear subspace. In this case, we consider the following spaces

$$
\begin{aligned}
Y^* &\subset \mathbb{P}_N^*, \text{ the dual variety of } Y \text{ in } \mathbb{P}_N, \\
L^* &\subset \mathbb{P}_N^*, \text{ the dual variety of } L \text{ in } \mathbb{P}_N, \\
L^\star & \qquad, \text{ the dual space of } L, \text{ i.e. the space of hyperplanes in } L, \text{ and} \\
Y_L^\star &\subset L^\star, \text{ the dual variety of } Y \text{ in } L.
\end{aligned}
$$

The inclusion $\widehat{L} \subset \mathbb{C}^{N+1}$ induces a linear map of the dual vector spaces

$$(\mathbb{C}^{N+1})^* \longrightarrow \widehat{L}^*,$$

and passing to the projective spaces we obtain a rational projection

$$\pi : \mathbb{P}_N^* \dashrightarrow L^\star$$

with center

$$L^* = \mathbb{P}\left(\{\varphi \in (\mathbb{C}^{N+1})^* : \varphi(L) = 0\}\right) \subset \mathbb{P}_N^*.$$

It is clear that

$$Y^* = \pi^{-1}(Y_L^\star),$$

and that this is a cone with vertex $L^* \subset Y^*$. As an illustration, the reader may take a look at a plane curve X in $\mathbb{P}_3$.

The relation between Y^* and $Y_L^\star$ obtained above for a degenerated variety can be dualized for a cone $X \subset \mathbb{P}_N$ with vertex E. Recall that we denote by

$$X^*, E^* \subset \mathbb{P}_N^* \quad \text{the dual varieties of } X \text{ and } E \text{ in } \mathbb{P}_N,$$

and we know $X^* \subset E^*$. The image of the cone $\widehat{X} \subset \mathbb{C}^{N+1}$ in $\mathbb{C}^{N+1}/\widehat{E}$ gives a projective variety

$$X_E \subset \mathbb{P}(\mathbb{C}^{N+1}/\widehat{E}).$$

With respect to this ambient space, there is a dual variety

$$X_E^\star \subset \mathbb{P}(\mathbb{C}^{N+1}/\widehat{E})^*,$$

and furthermore we have a canonical inclusion

$$\mathbb{P}(\mathbb{C}^{N+1}/\widehat{E})^* \subset \mathbb{P}_N^*.$$

By looking at the above definitions, it is easy to see that

$$X^* = X_E^\star \subset \mathbb{P}(\mathbb{C}^{N+1}/\widehat{E})^* = E^*;$$

i.e. X^* can be considered as the image of $X_E^\star$ under the inclusion map.

The interplay between cones and degenerate varieties via duality can be illustrated by the following scheme:

$$E = L^* \subset X = Y^* \subset \mathbb{P}_N,$$
$$Y = X^* \subset L = E^* \subset \mathbb{P}_N^*.$$

2.2.7 Tangent and Osculating Scrolls

In our next series of examples, we consider tangential varieties.

Example 1. Take a curve $C \subset \mathbb{P}_N$, i.e. a one-dimensional variety. We consider the rational map

$$\gamma : C \dashrightarrow \mathbb{G}(1, N) \quad \text{with} \quad \gamma(p) = \mathbb{T}_p C \text{ for smooth } p \,.$$

For a singular point $p \in C$, the image $\gamma(p)$ is finite. We define

$$B := \gamma(C) \subset \mathbb{G}(1, N) \quad \text{and} \quad \text{Tan}\, C := \bigcup_{L \in B} L \subset \mathbb{P}_N \,.$$

The variety $\text{Tan}\, C$ is called the *tangent scroll* of C. In case C is a line, $\text{Tan}\, C = C$. Otherwise, we have the

Proposition. *If $C \subset \mathbb{P}_N$ is a curve, but not a line, the tangent scroll $\text{Tan}\, C \subset \mathbb{P}_N$ is an irreducible surface and the canonical 1-dimensional almost ruling is developable.*

Proof. Everything is clear except for the developability. So take a smooth point $p \in C$ and a parameterization

$$\varphi : S \longrightarrow C \subset \mathbb{P}_N$$

of a neighborhood of p, where $o \in S \subset \mathbb{C}$ is an open neighborhood. We may furthermore assume that B is smooth in $\gamma(p)$ and

$$\gamma \circ \varphi : S \longrightarrow B$$

parameterizes a neighborhood of $\gamma(p)$. If

$$\varrho : S \longrightarrow \mathbb{C}^{N+1}$$

is a lifting of φ, then (ϱ, ϱ') is a local basis of $\gamma \circ \varphi$, which just means

$$\mathbb{T}_{\varphi(s)} C = \mathbb{P}(\text{span}\, \{\varrho(s), \varrho'(s)\}) \quad \text{for } s \in S \,.$$

Now the condition for developability is

$$\text{rank}\, (\varrho, \varrho', \varrho', \varrho'') = 3 \,.$$

Obviously, the rank is bounded by 3, and it cannot be smaller since $\dim \text{Tan}\, C = 2$. This can also be seen more directly: if the rank is 2 for every s, then ϱ'' would be contained in the plane spanned by ϱ and ϱ' for every s, so the curve $\gamma \circ \varphi : S \to \mathbb{G}(1, N)$ would have drill 0. By 1.5.2, it would be constant and C would be a line. $\square$

It can also be seen directly that the tangent plane to $\text{Tan}\, C$ is constant along a tangent line of C. We use the affine cone $\widehat{C} \subset \mathbb{C}^{N+1}$ and the holomorphic map ϱ as above. Then

$$S \times \mathbb{C} \longrightarrow \widehat{C}\,, \quad (s, \lambda) \longmapsto \lambda \varrho(s) \,,$$

is a local parameterization for $\lambda \neq 0$ which yields a local parameterization

$$\psi : S \times \mathbb{C}^2 \longrightarrow (\operatorname{Tan} C)^\wedge , \qquad (s, \lambda, \mu) \longmapsto \lambda \varrho(s) + \mu \varrho'(s) .$$

Hence, the tangent vector space of $(\operatorname{Tan} C)^\wedge$ at $\psi(s, \lambda, \mu)$ is

$$\operatorname{span} \{\varrho(s), \varrho'(s), \lambda \varrho'(s) + \mu \varrho''(s)\} = \operatorname{span} \{\varrho(s), \varrho'(s), \varrho''(s)\}$$

if $\mu \neq 0$. This shows the independence of the tangent space outside C. The fact that ψ is not immersive at $\mu = 0$ suggest that $\operatorname{Tan} C$ is singular in C. We will show later (3.3.3) that this is indeed the case.

Example 2. A *rational curve* $C \subset \mathbb{P}_N$ of *degree* $d \geq 1$ is the image of a regular map

$$\varphi : \mathbb{P}_1 \longrightarrow C \subset \mathbb{P}_N , \qquad (s_0 : s_1) \longmapsto (\varphi_0(s) : \ldots : \varphi_N(s)) ,$$

where $\varphi_0, \ldots, \varphi_N \in \mathbb{C}[s_0, s_1]$ are homogenous of degree d without a common zero in $\mathbb{P}_1$. We claim that the tangent scroll $\operatorname{Tan} C$ is the image of the rational map

$$\Phi : \mathbb{P}_1 \times \mathbb{P}_1 \quad \dashrightarrow \quad \operatorname{Tan} C \subset \mathbb{P}_N ,$$

$$(s_0 : s_1 ; t_0 : t_1) \longmapsto \left(t_0 \frac{\partial \varphi_0}{\partial s_0}(s) + t_1 \frac{\partial \varphi_0}{\partial s_1}(s) : \ldots : t_0 \frac{\partial \varphi_N}{\partial s_0}(s) + t_1 \frac{\partial \varphi_N}{\partial s_1}(s) \right) .$$

If φ is an immersion at $s = (s_0 : s_1)$, then the vectors $\frac{\partial \varphi}{\partial s_0}(s), \frac{\partial \varphi}{\partial s_1}(s) \in \mathbb{C}^{N+1}$ are linearly independent and

$$\Phi(\{s\} \times \mathbb{P}_1) = \mathbb{P}\left(\operatorname{span} \left\{ \frac{\partial \varphi}{\partial s_0}(s), \frac{\partial \varphi}{\partial s_1}(s) \right\} \right)$$

is a tangent line at $\varphi(s)$; hence, the assertion follows from the usual closure arguments. By EULER's formula, we have $C = \Phi(\Delta)$, where $\Delta \subset \mathbb{P}_1 \times \mathbb{P}_1$ is the diagonal.

The first associated curve

$$\varphi^{(1)} : \mathbb{P}_1 \longrightarrow \mathbb{G}(1, N)$$

is described by the Plücker coordinates of the tangent lines:

$$p_{ij} = \frac{\partial \varphi_i}{\partial s_0} \cdot \frac{\partial \varphi_j}{\partial s_1} - \frac{\partial \varphi_j}{\partial s_0} \cdot \frac{\partial \varphi_i}{\partial s_1} , \qquad 0 \leq i \leq j \leq N .$$

Finally, the *osculating scroll* $\operatorname{Tan}^{(k)} C$ of *order* k (see Example 3 below) is the image of the rational map

$$\Phi^{(k)} : \mathbb{P}_1 \times \mathbb{P}_k \quad \dashrightarrow \quad \operatorname{Tan}^{(k)} C \subset \mathbb{P}_N \quad \text{with}$$

$$\Phi^{(k)}(s_0 : s_1 ; t_0 : \ldots : t_k) = \sum_{j=0}^{k} \binom{k}{j} \frac{\partial^k \varphi}{\partial s_0^{k-j} \partial s_1^j} t_j .$$

Again, EULER's formula implies that $C = \Phi^{(k)}(\Delta_k)$, where

$$\Delta_k := \left\{ (s_0 : s_1 ; t_0 : \ldots : t_k) \in \mathbb{P}_1 \times \mathbb{P}_k : t_0 = s_0^k, \; t_1 = s_0^{k-1} s_1, \ldots, \; t_k = s_1^k \right\}$$

is the graph of the rational normal curve of degree k.

Example 3. Instead of the tangent line, we may take the osculating k-plane at every point of a curve; their union is the *osculating scroll* of *order k*. For a precise description, we use the *associated curves* (1.5.4). If

$$\varphi : S \longrightarrow C \subset \mathbb{P}_N$$

is a parameterization with a compact Riemann surface S, then for $0 \leq k \leq \operatorname{rank} \varphi$ we have an associated curve

$$\varphi^{(k)} : S \longrightarrow \mathbb{G}(k, N).$$

The space $\varphi^{(k)}(s)$ is the *osculating k-plane* of the curve $C = \varphi(S)$ at the point $p = \varphi(s)$. It is unique if p has only one inverse image in S; in this case,

$$\mathbb{T}_p^{(k)} C \in \mathbb{G}(k, N).$$

If p has more than one inverse image in S, there may be several (but only finitely many) different osculating k-planes.

General theory, for instance use REMMERT's mapping theorem and the theorem of CHOW [F$_1$, 1.18, 4.3], implies that the image

$$B_k := \varphi^{(k)}(S) \subset \mathbb{G}(k, N)$$

is algebraic; hence,

$$\operatorname{Tan}^{(k)} C := \bigcup_{E \in B_k} E \subset \mathbb{P}_N$$

is a variety. It is the closure of the union of all osculating k-planes and called the *osculating scroll* of *order k*.

Now it is easy to see that the canonical ruling of Tan $^{(k)}C$ is developable. We take a coordinate neighborhood $o \in U \subset S$ and a lifting

$$\varrho : U \longrightarrow \mathbb{C}^{N+1}$$

of φ such that $(\varrho(s), \varrho'(s), \dots, \varrho^{(k)}(s))$ are linearly independent for all $s \in U$. Then

$$\varrho, \varrho', \dots, \varrho^{(k)} : U \longrightarrow \mathbb{C}^{N+1}$$

is a local basis of $\varphi^{(k)}$ on U. Now obviously,

$$\operatorname{rank} \left(\varrho, \varrho', \dots, \varrho^{(k)}, \varrho', \varrho'', \dots, \varrho^{(k+1)} \right) = \operatorname{rank} \left(\varrho, \varrho', \dots, \varrho^{(k+1)} \right) \leq k + 2$$

on U. If $k = \operatorname{rank} \varphi$, then $\varphi^{(k)}$ is constant. Otherwise, we can choose U such that $\varrho^{(k+1)}$ is linearly independent of $\varrho, \dots, \varrho^{(k)}$ on U. These steps yield the proof of the following proposition.

Proposition. *If $C \subset \mathbb{P}_N$ is a curve, then for $0 \le k < \operatorname{rank} C$ the osculating scroll $\operatorname{Tan}^{(k)} C \subset \mathbb{P}_N$ is a variety of dimension $k + 1$ and the canonical k-dimensional uniruling is developable.*

If we return to the map

$$\varphi : S \longrightarrow C \,,$$

where S is a compact Riemann surface, then the above result means that the tangent space of $\operatorname{Tan}^{(k)} C$ is constant along smooth points of the linear space $\varphi^{(k)}(s)$ for $s \in S$. In fact, for every such $p \in \varphi^{(k)}(s) \subset \operatorname{Tan}^{(k)} C$ we have

$$\mathbb{T}_p(\operatorname{Tan}^{(k)} C) = \varphi^{(k+1)}(s) \,;$$

this can be proved inductively as in Example 1. In the same way, $C \subset \operatorname{Sing}(\operatorname{Tan} C)$ implies

$$\operatorname{Tan}^{(k-1)} C \subset \operatorname{Sing}(\operatorname{Tan}^{(k)} C) \,.$$

The above Proposition is also a direct consequence of Lemma 2.2.8.

2.2.8 Classification of Developable One Parameter Rulings

As we have seen in the previous sections, obvious examples for developable ruled varieties are cones and osculating scrolls of curves. In this paragraph, we use the Normal Form Lemma of 1.5.5 to show the following theorem.

Classification Theorem. *Every developable uniruled variety $X \subset \mathbb{P}_N$ with a one-dimensional base $B \subset \mathbb{G}(k, N)$ is a cone over an osculating scroll of some curve $C \subset \mathbb{P}_N$.*

More explicitly: if $B \subset \mathbb{G}(k, N)$ is a one-dimensional variety and if the ruling of

$$X := \bigcup_{L \in B} L \subset \mathbb{P}_N$$

is developable, then there is a curve $C \subset \mathbb{P}_N$, an integer l, and a linear subspace $E \subset \mathbb{P}_N$ such that

$$X = \operatorname{Tan}^{(l)} C \,\#\, E \,.$$

In particular, $\dim X = k + 1 = \dim E + l + 1$.

The drill of a curve

$$\psi : S \longrightarrow \mathbb{G}(k, N)$$

was measured in 1.5.2 by a number $d(\psi)$ with $0 \le d(\psi) \le k + 1$. Recall that $d(\psi) = 0$ is equivalent to $\psi = \text{constant}$. Using this number, we can formulate the

Lemma. *If S is a compact Riemann surface, $\psi : S \to \mathbb{G}(k, N)$ is a nonconstant holomorphic map and*

$$X := \bigcup_{s \in S} \psi(s) \subset \mathbb{P}_N \,,$$

then the following conditions are equivalent:

i) the k-dimensional uniruling of X, defined by ψ, is developable,

ii) $d(\psi) = 1$.

Proof of the Classification Theorem. Take a desingularisation

$$\psi : S \longrightarrow B \subset \mathbb{G}(k, N).$$

Then the above lemma implies $d(\psi) = 1$, and by the Normal Form Lemma 1.5.5, we have

$$\psi = \varphi^{(l)} \oplus V,$$

where $V \subset \mathbb{C}^{N+1}$ is a vector space, $\varphi : S \to \mathbb{P}_N$ is a curve, and l is an integer. This permits us to define

$$C := \varphi(S) \quad \text{and} \quad E := \mathbb{P}(V). \qquad \qquad \Box$$

Proof of the lemma. We use an open set $U \subset S$ and a local basis

$$\varrho_0, \ldots, \varrho_k : U \longrightarrow \mathbb{C}^{N+1}$$

of ψ on U. By the definition of $d(\psi)$, we have

$$\text{rank}\,(\varrho_0, \ldots, \varrho_k, \varrho_0', \ldots, \varrho_k') = k + 1 + d(\psi)$$

almost everywhere. The developability criterion in 2.2.4 is

$$\text{rank}\,(\varrho_0, \ldots, \varrho_k, \varrho_0', \ldots, \varrho_k') = k + 2$$

almost everywhere; this implies the equivalence. $\qquad \qquad \Box$

2.2.9 Example of a "Twisted Plane"

The only obvious examples of developable almost rulings with a base of dimension bigger than one are cones. A more sophisticated example with two-dimensional base was suggested by BOURGAIN [W]. It turned out [AG$_1$] that this is the algebraic version of an older example of SACKSTEDER [S].

We will consider the variety

$$E := \{(x; s) \in \mathbb{P}_4 \times \mathbb{P}_1 : s_1 x_0 - s_0 x_4 = s_0^2 x_1 + s_0 s_1 x_2 + s_1^2 x_3 = 0\} \subset \mathbb{P}_4 \times \mathbb{P}_1.$$

For every $s \in \mathbb{P}_1$, the fiber $E_s \subset \mathbb{P}_4$ of the second projection is a plane. The image of E under the first projection is the hypersurface

$$X = \bigcup_{s \in \mathbb{P}_1} E_s = \{x \in \mathbb{P}_4 : x_0^2 x_1 + x_0 x_4 x_2 + x_4^2 x_3 = 0\} \subset \mathbb{P}_4,$$

where the equation is obtained by elimination of s_0, s_1. This hypersurface has an almost ruling by planes with a one-dimensional base $B \subset \mathbb{G}(2, 4)$, which is the image of the map

$$\mathbb{P}_1 \xrightarrow{\varphi} \mathbb{G}^*(1, 4) \xrightarrow{D} \mathbb{G}(2, 4),$$

where $\mathcal{D}$ is duality, and φ is given by the 2-minors of the matrix

$$\begin{pmatrix} s_1 & 0 & 0 & 0 & -s_0 \\ 0 & s_0^2 & s_0 s_1 & s_1^2 & 0 \end{pmatrix}.$$

The hypersurface $X \subset \mathbb{P}_4$ is interesting for several reasons. By computing the gradient of the defining polynomial, it is easy to see that

$$\text{Sing } X = \{x \in \mathbb{P}_4 : x_0 = x_4 = 0\}.$$

This is a plane at infinity; in particular, the affine part of X is smooth.

The geometry of X is easy to describe. The maps

$$\lambda : \mathbb{P}_1 \longrightarrow \mathbb{P}_4, \ (s_0 : s_1) \longmapsto (s_0 : 0 : 0 : 0 : s_1), \quad \text{and}$$

$$\beta : \mathbb{P}_1 \longrightarrow \mathbb{P}_4, \ (s_0 : s_1) \longmapsto (0 : s_1^2 : -2s_0 s_1 : s_0^2 : 0)$$

parameterize a line and a conic $C \subset \text{Sing } X \subset \mathbb{P}_4$. The tangent line $\mathbb{T}_s C$ to C at $\beta(s)$ has equations

$$x_0 = s_0^2 x_1 + s_0 s_1 x_2 + s_1^2 x_3 = x_4 = 0;$$

hence,

$$E_s = \mathbb{T}_s C \# \lambda(s).$$

This explains the name *twisted plane* for the hypersurface X. A generalization was considered in [V].

A geometric explanation for the arising of singularities of X is the following. The planes E_s are disjoint outside Sing X, and every point $p \in \text{Sing } X \backslash C$ is contained in precisely the two planes containing the two tangents of C passing through p (see Example 2 of 2.2.1).

On the affine part $S = \{(s_0 : s_1) \in \mathbb{P}_1 : s_0 \neq 0\}$, we have the coordinate $s = \frac{s_1}{s_0}$ and a local basis (1.5.1)

$$\varrho_0 = (1, 0, 0, 0, s),$$
$$\varrho_1 = (0, s, -1, 0, 0),$$
$$\varrho_2 = (0, s^2, -2s, 1, 0).$$

Obviously $\varrho_2' = 2\varrho_1$, and rank $(\varrho_0, \varrho_1, \varrho_2, \varrho_0', \varrho_1', \varrho_2') = 5$ is valid everywhere on S; hence, the curve $B \subset \mathbb{G}(2, 4)$ has drill $d = 2$, and the almost ruling by planes of X is not developable.

However, we can construct a developable almost ruling by lines $B_1 \subset \mathbb{G}(1, 4)$ of the above hypersurface X. Let

$$X' = \{(x_1, x_2, x_3, x_4) \in \mathbb{C}^4 : x_1 + x_4 x_2 + x_4^2 x_3 = 0\}$$

be the affine part of X. Then every affine plane $E_s' = E_s \cap X'$ can be ruled by parallel lines in the direction of the point $\beta(s)$ at infinity, i.e. in the direction of

$$(s^2, -2s, 1, 0) \in \mathbb{C}^4.$$

We obtain a parameterization

$$\Phi : \mathbb{C}^3 \longrightarrow X' \subset \mathbb{C}^4 , \ (s_1, s_2, t) \longmapsto \alpha(s) + t\varrho(s) ,$$

$$\alpha(s) := (s_1 s_2, -s_2, 0, s_1) , \ \varrho(s) := (s_1^2, -2s_1, 1, 0) .$$

This ruling of X' by lines extends to X. To obtain the base $B_1 \subset \mathbb{G}(1, 4)$, we consider the rational map

$$\psi : \mathbb{P}_2 \dashrightarrow \mathbb{G}(1, 4) \subset \mathbb{P}_9 ,$$

where $\psi(s_0 : s_1 : s_2)$ is spanned by the rows of the matrix

$$\begin{pmatrix} s_0^2 & s_1 s_2 & -s_0 s_2 & 0 & s_0 s_1 \\ 0 & s_1^2 & -2s_0 s_1 & s_0^2 & 0 \end{pmatrix} .$$

The map ψ has base points on the line $s_0 = 0$. Now, our base is the rational surface

$$B_1 := \psi(\mathbb{P}_2) \subset \mathbb{G}(1, 4) .$$

In order to show that this 1-dimensional almost ruling of X is developable, we put $s_0 = 1$ and obtain the local basis

$$\varrho_0 = (1, s_1 s_2 , -s_2, 0, s_1)$$
$$\varrho_1 = (0, s_1^2 , -2s_1, 1, 0) .$$

A trivial computation shows that

$$\frac{\partial \varrho_1}{\partial s_1} = 2\frac{\partial \varrho_0}{\partial s_2} , \quad \frac{\partial \varrho_1}{\partial s_2} = 0 , \quad \text{and} \quad \text{rank} \left(\varrho_i, \frac{\partial \varrho_i}{\partial s_j} \right) = 4 .$$

The affine part X' is not a cylinder, since ϱ and $\frac{\partial \varrho}{\partial s_1}$ are linearly independent; hence, X is not a cone. More precisely: all the lines of the ruling contained in the same plane E_s meet at infinity on the conic C in the point $\beta(s)$, which is different for different values of $s = (s_0 : s_1)$.

Now we take a look at the dual variety $X^* \subset \mathbb{P}_4^*$. To describe X^* explicitly, we use the defining polynomial

$$F = x_0^2 x_1 + x_0 x_2 x_4 + x_3 x_4^2$$

of X. Since X is a hypersurface, the dual variety is the image of the rational map

$$\delta : X \dashrightarrow X^* \subset \mathbb{P}_4^*$$

defined by the gradient

$$\text{grad } F = (2x_0 x_1 + x_2 x_4, x_0^2, x_0 x_4, x_4^2, x_0 x_2 + 2x_3 x_4) .$$

The map δ restricted to the line $(s_0 : 0 : 0 : 0 : s_1)$ yields the conic

$$\mathbb{P}_1 \longrightarrow X^* , \quad (s_0 : s_1) \longmapsto (0 : s_0^2 : s_0 s_1 : s_1^2 : 0) .$$

For $s \in \mathbb{P}$, we looked at the plane E_s. The map δ is constant on all lines passing $\beta(s) \in C$. To obtain the values of δ, we choose a point

$$v = (0 : v_1 : v_2 : v_3 : 0)$$

on the tangent $\mathbb{T}_s C$ to the conic C, but different from $\beta(s)$. This choice implies

$$\frac{2s_0 v_1 + s_1 v_2}{s_0 v_2 + 2s_1 v_3} = \frac{-s_1}{s_0} . \qquad (*)$$

Since X is singular at v, we approach this point by the line

$$\eta : \mathbb{P}_1 \longrightarrow X, \quad (\mu_0 : \mu_1) \longmapsto \mu_0(s_0 : 0 : 0 : 0 : s_1) + \mu_1 v .$$

A simple computation using $(*)$ shows

$$\delta(\eta(\mu)) = \mu_0(0 : s_0^2 : s_0 s_1 : s_1^2 : 0) + \mu_1(s_1 : 0 : 0 : 0 : -s_0) .$$

So X^* is ruled by lines; more precisely,

$$X^* = \bigcup_{s \in \mathbb{P}_1} \overline{(0 : s_0^2 : s_0 s_1 : s_1^2 : 0) (s_1 : 0 : 0 : 0 : -s_0)} \subset \mathbb{P}_4^*$$

is a ruled surface directed by a conic and a line. It is easy to see, that this ruling is not developable.

There is a more abstract way of describing X^* via the duality

$$\mathcal{D} : \mathbb{G}^*(1, 4) \longrightarrow \mathbb{G}(2, 4) .$$

The tangent hyperplanes to smooth points of X along the plane $E_s \subset X$ and their limits are all the hyperplanes $H \supset E_s$; this is just the line $\varphi(s) \subset \mathbb{P}_4^*$, where $\varphi(s) \in \mathbb{G}^*(1, 4)$ was defined above. Hence,

$$X^* = \bigcup_{s \in \mathbb{P}_1} \varphi(s) \subset \mathbb{P}_4^*, \quad \text{while} \quad X = \bigcup_{s \in \mathbb{P}_1} \mathcal{D}(\varphi(s)) \subset \mathbb{P}_4 .$$

Finally, let us restate the properties of the *twisted plane* $X \subset \mathbb{P}_4$:

- X has a nondevelopable ruling by planes

- X has a developable subruling by lines

- two lines of the subruling in the same plane meet at infinity

- Sing X is a plane at infinity

- the dual $X^* \subset \mathbb{P}_4^*$ is a nondevelopable ruled surface.

Evidently, the twisted plane is controlled by a line and a plane conic. This construction will be generalized in 2.3.5.

2.2.10 Characterization of Drill One Curves

It is clear that the developability of a uniruling has something to do with the tangents of the base in the Grassmannian. Basically, they have to satisfy certain restrictions. The best result can be obtained in case the base is a curve. Then the restriction can be expressed in terms of the drill of the curve introduced in 1.5.1.

Theorem. *If S is a compact Riemann surface and*

$$\psi : S \longrightarrow \mathbb{G}(k, N)$$

is a nonconstant holomorphic curve with $B := \psi(S)$, then the following conditions are equivalent:

i) $d(\psi) = 1$;

ii) if $E \in B$ is a smooth point, then $\mathbb{T}_E B \subset \mathbb{G}(k, N)$.

Here $\mathbb{T}_E B$ denotes the tangent line to B at E in the ambient projective space $\mathbb{P}(\bigwedge^{k+1} \mathbb{C}^{N+1})$.

Together with Lemma 2.2.8 this implies the

Corollary. *Assume $B \subset \mathbb{G}(k, N)$ is a curve and*

$$X := \bigcup_{E \in B} E \subset \mathbb{P}_N$$

is not linear and of dimension $k + 1$. Then the following conditions are equivalent:

i) the uniruling of X given by B is developable;

ii) if $E \in B$ is a smooth point then $\mathbb{T}_E B \subset \mathbb{G}(k, N)$.

Proof of the theorem. For $i) \Rightarrow ii)$ we choose a point $o \in S$ such that B is smooth at $E = \psi(o)$ and a local basis

$$\varrho_0, \ldots, \varrho_k : U \longrightarrow \mathbb{C}^{N+1}$$

of ψ in a neighborhood $o \in U \subset S$. If

$$\varrho := \varrho_0 \wedge \ldots \wedge \varrho_k : U \longrightarrow \bigwedge^{k+1} \mathbb{C}^{N+1} ,$$

then for every $s \in U$ we have

$$\varrho'(s) = \sum_{i=0}^{k} \varrho_0(s) \wedge \ldots \wedge \varrho_i'(s) \wedge \ldots \wedge \varrho_k(s) .$$

If we denote $v_i := \varrho_i(o)$ and $v_i' := \varrho_i'(o)$ for $i = 0, \ldots, k$, then $d(\psi) = 1$ implies that we may assume $o \in S$ and labeling are chosen such that

$$(v_0, \ldots, v_k, v_0') \quad \text{is a basis of span}\{v_0, \ldots, v_k, v_0', \ldots, v_k'\} =: V.$$

If $w := \varrho(o)$ and $w' := \varrho'(o)$, then $\mathbb{T}_E B \subset \mathbb{G}(k, N)$ is equivalent to

$$\text{span}\{w, w'\} \subset \text{Gr}(k+1, N+1).$$

By the Wedge Criterion in 1.2.4, this means

$$\dim\left\{v \in \mathbb{C}^{N+1} : (\lambda w + \mu w') \wedge v = 0 \in \bigwedge\nolimits^{k+2}\mathbb{C}^{N+1}\right\} \geq k+1$$

for every $(\lambda, \mu) \in \mathbb{C}^2$. If $v \notin V$, then obviously $(\lambda w + \mu w') \wedge v \neq 0$. Thus we may assume

$$v = \sum_{j=0}^{k} \alpha_j v_j + \alpha v_0' \in V.$$

Consider the linear form $F_{\lambda,\mu} \in V^*$ uniquely determined by

$$(\lambda w + \mu w') \wedge v = F_{\lambda,\mu}(\alpha_0, \ldots, \alpha_k, \alpha)\, v_0 \wedge \ldots \wedge v_k \wedge v_0'.$$

Since $\dim V = k+2$, we have $\dim \text{Ker}\, F_{\lambda,\mu} \geq k+1$ for every (λ, μ) and this implies $ii)$.

For the proof of $ii) \Rightarrow i)$, we use the above notations, but assume that $d := d(\psi) \geq 2$. Then $\dim V = k + d + 1$, and we may assume

$$(v_0, \ldots, v_k, v_0', \ldots, v_{d-1}') \quad \text{is a basis of } V.$$

Using the Wedge Criterion again, we see that if

$$K := \left\{v \in Y : w' \wedge v = 0\right\},$$

then it is sufficient to prove $\dim K < k+1$. This requires some computation. We have

$$v_j' = \sum_{i=0}^{k} a_{ij} v_i + \sum_{i=0}^{d-1} b_{ij} v_i' \quad \text{for } j = d, \ldots, k.$$

Using this, we can compute

$$w' = \sum_{j=0}^{d-1} v_0 \wedge \ldots \wedge v_{j-1} \wedge v_j' \wedge v_{j+1} \wedge \ldots \wedge v_k$$

$$+ a(v_0 \wedge \ldots \wedge v_k) + \sum_{j=d}^{k} \sum_{i=0}^{d-1} b_{ij}(v_0 \wedge \ldots \wedge v_{j-1} \wedge v_i' \wedge v_{j+1} \wedge \ldots \wedge v_k),$$

where $a := \sum_{j=d}^{k} a_{jj}$. The second factor for the wedge is written as

$$v = \sum_{j=0}^{k} \alpha_j v_j + \sum_{i=0}^{d-1} \beta_i v_i' .$$

We do not compute $w' \wedge v$ completely; we simply give the coefficients of some vectors in a basis of $\bigwedge^{k+1} V$:

$$(a\beta_j - \alpha_j)(v_0 \wedge \ldots \wedge v_k \wedge v_j') \quad \text{for } j = 0, \ldots, d-1 ,$$

$$(\beta_1 - \beta_0)(v_0' \wedge v_1 \wedge \ldots \wedge v_k \wedge v_1') .$$

In particular, these coefficients have to vanish, yielding $d + 1$ independent conditions for the coefficients α_j and β_i; hence,

$$\dim K \le (k + d + 1) - (d + 1) = k . \qquad \square$$

If the base of the ruling has arbitrary dimension, then the part $ii) \Rightarrow i)$ of the above theorem is an immediate consequence and the part $i) \Rightarrow ii)$ is false.

Corollary. *If the base $B \subset \mathbb{G}(k, N)$ of the uniruling of a variety $X \subset \mathbb{P}_N$ has the property that $\mathbb{T}_E B \subset \mathbb{G}(k, N)$ for every smooth point $E \in B$, then the uniruling is developable.*

Proof. Take an open set $o \in S \subset \mathbb{C}^r$, where $r = \dim B$, a parameterization

$$\psi : S \longrightarrow B \subset \mathbb{G}(k, N) , \quad \psi(o) = E ,$$

and a local basis

$$\varrho_0, \ldots, \varrho_k : S \longrightarrow \mathbb{C}^{N+1}$$

of ψ. We have to show

$$\text{rank} \left(\varrho_i, \frac{\partial \varrho_i}{\partial s_j} \right)_{i,j} \le k + r + 1 .$$

In S, we take lines S_j given by $s_l = 0$ for $l \ne j$. Then the curves $B_j = \psi(S_j) \subset B$ have the property

$$\mathbb{T}_E B_j \subset \mathbb{T}_E B \subset \mathbb{G}(k, N)$$

and the theorem implies

$$\text{rank} \left(\varrho_i, \frac{\partial \varrho_i}{\partial s_j} \right)_i \le k + 2 \quad \text{for every } j = 1, \ldots, r .$$

This means that for fixed j, the $(k+1)$-derivatives can increase the rank only by one. Thus, all together, they can increase it only by r. $\qquad \square$

Example. We use the hypersurface $X \subset \mathbb{P}_4$ from 2.2.9 with its developable ruling B_1 given by the rational map

$$\psi : \mathbb{P}_2 \dashrightarrow B_1 \subset \mathbb{G}(1, 4) .$$

If we restrict ψ to the line $\mathbb{P}_1 \subset \mathbb{P}_2$ with $s_2 = 0$, we obtain a curve

$$\widetilde{B}_1 = \psi(\mathbb{P}_1) \quad \text{and} \quad \widetilde{X} := \bigcup_{L \in \widetilde{B}_1} L \subset \mathbb{P}_4$$

is a surface ruled by lines. If we put $s_0 = 1, s_1 = s$, we have a local basis

$$\varrho_0 = (1, s, 0, 0, s) , \quad \varrho_1 = (0, s^2, -2s, 1, 0)$$

and rank $(\varrho_0, \varrho_1, \varrho_0', \varrho_1') = 4$, hence $\widetilde{X}$ is not developable.

This implies that we cannot have

$$\mathbb{T}_E B_1 \subset \mathbb{G}(1, 4)$$

for every smooth $E \in B_1$; otherwise we would have $\mathbb{T}_E \widetilde{B}_1 \subset \mathbb{G}(1, 4)$, and by the above Corollary $\widetilde{X}$ would be developable.

With some computation, it can be shown directly that $\mathbb{T}_E B_1 \not\subset \mathbb{G}(1, 4)$.

2.3   The Gauß Map

A given variety may have several developable rulings of different dimensions. The ruling of highest possible dimension is the most important one, and it can be detected via the fibration determined by the Gauß map. In classical differential geometry, the Gauß map is defined for a surface in $\mathbb{R}^3$ and has values in the 2-sphere (0.1). The adequate generalization has values in a Grassmannian.

2.3.1 Definition of the Gauß Map

In the simplest case, $X \subset \mathbb{P}_N$ is a smooth hypersurface. If F is a defining homogeneous polynomial for X, then for every $p \in X$ we can choose a $q \in \widehat{X} \backslash \{0\}$ lying over p, and we obtain a well defined regular map

$$\gamma^* : X \longrightarrow \mathbb{P}_N^* = \mathbb{G}^*(0, N) , \quad p \longmapsto \left(\frac{\partial F}{\partial x_0}(q) : \ldots : \frac{\partial F}{\partial x_N}(q) \right) .$$

$\gamma^*(p)$ is just the dual of the tangent hyperplane $\mathbb{T}_p X \subset \mathbb{P}_N$ of X at p. The dual regular map

$$\gamma : X \longrightarrow \mathbb{P}_N^\vee = \mathbb{G}(N - 1, N) , \quad p \longmapsto \mathbb{T}_p X ,$$

is called the *Gauß map* of the hypersurface X.

Example. If $X \subset \mathbb{P}_N$ is a smooth quadric, we use the isomorphism

$$\alpha : \mathbb{P}_N \longrightarrow \mathbb{P}_N^* \quad \text{with } \alpha(X) = X^*$$

from Example 2.1.4. The Gauß map $\gamma : X \to X^* \subset \mathbb{P}_N^*$ is just the restriction of α.

If $X \subset \mathbb{P}_N$ is a smooth variety of arbitrary dimension n, we have $X = V(F_1, \ldots, F_m)$ with homogeneous polynomials F_i. For a $q \in \mathbb{C}^{N+1}$, we put

$$\operatorname{grad}_q F := \left(\frac{\partial F}{\partial x_0}(q), \ldots, \frac{\partial F}{\partial x_N}(q) \right) \in \left(\mathbb{C}^{N+1} \right)^* .$$

By choosing q over $p \in X$, we obtain a regular map

$$\gamma^* : X \longrightarrow \mathbb{G}^*(N - n - 1, N), \quad p \longmapsto \mathbb{P}\left(\operatorname{span}\{\operatorname{grad}_q F_1, \ldots, \operatorname{grad}_q F_m\} \right),$$

and again the dual of $\gamma^*(p)$ is $\mathbb{T}_p X$ (1.1.1); hence, the *Gauß map*

$$\gamma : X \longrightarrow \mathbb{G}(n, N), \quad p \longmapsto \mathbb{T}_p X,$$

is regular.

In case that $X = V(F_1, \ldots, F_m)$ is irreducible, n-dimensional, and possibly singular, we define, with the notations as above

$$r(p) := \operatorname{rank}\left(\operatorname{grad}_q F_1, \ldots, \operatorname{grad}_q F_m \right) .$$

By 1.1.1, we know $r(p) \le N - n$, with equality iff p is smooth. The Extension Lemma in 1.2.6 implies that there is a unique rational map

$$\gamma^* : X \dashrightarrow \mathbb{G}^*(N - n - 1, N) \quad \text{with} \quad p \longmapsto \mathbb{P}\left(\operatorname{span}\{\operatorname{grad}_q F_1, \ldots, \operatorname{grad}_q F_m\} \right)$$

for every smooth $p \in X$. By duality, we obtain the rational *Gauß map*

$$\gamma : X \dashrightarrow \mathbb{G}(n, N), \quad p \longmapsto \mathbb{T}_p X \quad \text{for smooth } p .$$

Geometrically, it is not at all obvious what happens precisely in a singular point p: the rank $r(p)$ drops down, but $\gamma^*(p)$ may be more than one point; the dimension of $\mathbb{T}_p X = \mathbb{P}(T_q \widehat{X})$ jumps up, and — this sounds more reasonable — $\gamma(p)$ may consist of more than one n-dimensional linear subspace. But for many arguments it is sufficient to know that the Gauß map just exists as a rational map, where the critical images $\gamma(p)$ are as small as possible.

2.3.2 Linearity of the Fibers

Let us now take a look at the image and the fibers of the Gauß map

$$\gamma : X \dashrightarrow \mathbb{G}(n, N)$$

for a variety $X \subset \mathbb{P}_N$ with $\dim X = n$. The rational map γ is given by its graph

$$\Gamma_\gamma \subset X \times \mathbb{G}(n, N),$$

which is an irreducible algebraic set of dimension n. Denote by

$$\pi : \Gamma_\gamma \longrightarrow X \quad \text{and} \quad \xi : \Gamma_\gamma \longrightarrow \mathbb{G}(n, N)$$

the canonical projections, then for a smooth point $p \in X$ we have

$$\pi^{-1}(p) = (p, \mathbb{T}_p X).$$

The *Gauß image* of X is defined as $\gamma(X) := \xi(\Gamma_\gamma) \subset \mathbb{G}(n, N)$, and its dimension

$$r(X) := \dim \gamma(X)$$

is called the *Gauß rank* of X. If $E \in \mathbb{G}(n, N)$, then

$$\gamma^{-1}(E) := \pi\left(\xi^{-1}(E)\right) = \pi\left(\Gamma_\gamma \cap X \times \{E\}\right) \subset X$$

is the *Gauß fiber* over E.

Now take $E \in \operatorname{Im} \gamma$. We consider the dual variety $X^\vee \subset \mathbb{P}_N^\vee$ (2.1.4) and the linear space

$$E^\vee := \left\{H \in \mathbb{P}_N^\vee : H \supset E\right\} \subset X^\vee.$$

Unless $E^\vee \subset \operatorname{Sing} X^\vee$, we can find $H_1, \ldots, H_m \in X_{\mathrm{sm}}^\vee$, $m = N - n$ such that

$$E^\vee = \operatorname{span}\{H_1, \ldots, H_m\} \quad \text{and} \quad E = H_1 \cap \ldots \cap H_m.$$

By 2.1.6, the contact loci

$$X_{H_i} = \text{closure of } \{x \in X_{\mathrm{sm}} : \mathbb{T}_x X \subset H_i\} \subset X$$

are linear and obviously

$$\gamma^{-1}(E) = X_{H_1} \cap \ldots \cap X_{H_m}.$$

Hence $\gamma^{-1}(E)$ is linear. Note that

$$\gamma^{-1}(E) \cap X_{\mathrm{sm}} = \{x \in X_{\mathrm{sm}} : \mathbb{T}_x X = E\} \subset E,$$

since $x \in \mathbb{T}_x X$. This implies

$$\gamma^{-1}(\gamma(p)) \subset \mathbb{T}_p X$$

for generic $p \in X$. The dimension of the fiber of the rational map γ can be computed via the dimension formula for the regular map ξ. Thus we have the basic

Linearity Theorem. *If $X \subset \mathbb{P}_N$ is a variety of dimension n with Gauß map*

$$\gamma : X \dashrightarrow \mathbb{G}(n, N)$$

and Gauß rank $r(X) = \dim \gamma(X)$, then the generic fiber $\gamma^{-1}(\gamma(p)) \subset \mathbb{T}_p X$ is a linear subspace of dimension $n - r(X)$.

A nontrivial consequence of the Linearity Theorem is the connectedness of the general fiber. By the above argument the fiber $\gamma^{-1}(E)$ is generic, if $E^\vee$ is not contained in the singular locus of $X^\vee$. This condition is not so easy to check; in 2.4.6, we will give a different local approach to this linearity result. In case of a hypersurface X, a very elementary proof can be given as in 0.2 for a surface in $\mathbb{R}^3$ [FW, Appendix].

2.3.3 Gauß Map and Developability

If a projective variety $X \subset \mathbb{P}_N$ has dimension n, its Gauß image $\gamma(X)$ dimension r, and $k := n - r$, then by the above Linearity Theorem we have

$$\kappa(p) := \gamma^{-1}(\gamma(p)) \in \mathbb{G}(k, N)$$

for almost every $p \in X$. In 2.4.3, we can give a simple proof that this yields a rational map

$$\kappa : X \dashrightarrow \mathbb{G}(k, N) .$$

This implies that the image $\kappa(X)$ determines an almost ruling of X in the sense of 2.2.1, and by definition of the Gauß map it is developable. It is maximal in the sense that the dimension k of the linear leaves is maximal. A dimension count implies that

$$\dim \kappa(X) = \dim \gamma(X) ;$$

hence, $\kappa(X)$ is a base of minimal dimension for a developable uniruling of X.

Generically, the Gauß rank is equal to the dimension; any developable uniruling of positive dimension is exceptional and, in particular, produces singularities. This follows from a

Theorem of ZAK. *If the variety $X \subset \mathbb{P}_N$ is smooth, then the Gauß map $\gamma : X \to \mathbb{G}(n, N)$ is regular, finite, and birational. In particular, the generic fiber is a single point.*

A proof of this theorem will be given in 3.1.3. Of course not every fiber has to be connected; for instance, a plane curve has a finite number of double tangents.

If there is a developable uniruling with 1-dimensional base, then the Gauß rank is at most 1. Hence, the Classification Theorem proved in 2.2.8 also classifies varieties of Gauß rank 1. The case of hypersurfaces of Gauß rank 2 will be considered in 2.5.6.

2.3.4 Gauß Image and Dual Variety

We will now explore on an interesting relation between the Gauß image and the dual variety. We will use $\mathbb{P}_N^\vee = \mathbb{G}(N - 1, N)$, the incidence variety

$$I = \left\{ (E, H) \in \mathbb{G}(n, N) \times \mathbb{P}_N^\vee : E \subset H \right\} ,$$

which is algebraic by Proposition 2.1.1, and the canonical projections

$$\mathbb{G}(n, N) \xleftarrow{\pi} I \xrightarrow{\xi} \mathbb{P}_N^\vee .$$

For an n-dimensional variety $X \subset \mathbb{P}_N$, we have the Gauß image $\gamma(X) \subset \mathbb{G}(n, N)$ and the dual variety $X^\vee \subset \mathbb{P}_N^\vee$. From the definition of these two varieties, it follows immediately that

$$X^\vee = \xi \left(\pi^{-1}(\gamma(X)) \right) ,$$

and the Duality Theorem 2.1.5 implies

$$X = \left(\xi \left(\pi^{-1}(\gamma(X)) \right) \right)^{\vee} .$$

Thus, we have obtained a method of reconstructing X from its Gauß image $\gamma(X)$. It is much simpler in the case that X is a hypersurface, then

$$\gamma(X) = X^{\vee} \quad \text{and} \quad X = \gamma(X)^{\vee} .$$

The incidence variety considered above can be used for another comparison of the Gauß image and the dual variety. In 2.1.4, we had defined the dual defect $\delta_*(X)$. A similar measure is the *Gauß defect*

$$\delta_{\gamma}(X) := n - \dim \gamma(X) .$$

Proposition. *For every variety $X \subset \mathbb{P}_N$ of dimension n we have*

$$0 \le \delta_{\gamma}(X) \le \delta_*(X) \le n .$$

Proof. As above, consider

$$I_X := \pi^{-1}(\gamma(X)) \subset I \subset \mathbb{G}(n, N) \times \mathbb{P}_N^{\vee} .$$

Since the generic fiber of I_X over $\gamma(X)$ is linear of dimension $N - n - 1$, we have

$$\dim X^* \le \dim I_X = \dim \gamma(X) + N - n - 1 ; \quad \text{hence,}$$

$$n - \dim \gamma(X) \le N - 1 - \dim X^* . \qquad \square$$

2.3.5 Existence of Varieties with Given Gauß Rank

As we have seen in 2.3.2 and 2.3.3, a variety $X \subset \mathbb{P}_N$ with $n = \dim X$ has a Gauß rank $r = \dim \gamma(X)$ and

$$k := n - r$$

is the dimension of the general linear fiber of the Gauß map γ. For a general X, we have $r = n$. Under the only obvious assumption

$$0 \le r < n < N ,$$

all combinations of r and n can occur:

Trivial examples are cones. Take some variety $Y \subset \mathbb{P}_N$ such that

$$\dim Y = \dim \gamma(Y) = r ,$$

and a disjoint linear space $E \subset \mathbb{P}_N$ with

$$\dim E = n - r - 1 = k - 1 .$$

Then the cone $X := Y \# E$ has dimension n and Gauß rank r (Corollary 2.2.5).

Examples which are not cones (2.2.6) are more difficult to find. One method is to generalize the construction of the twisted plane $X \subset \mathbb{P}_4$ of dimension 3 (2.2.9) to a *twisted* $(n-1)$-*plane* $X \subset \mathbb{P}_N$ of dimension n.

We start with a curve in the Grassmannian of $(n-1)$-planes in $\mathbb{P}_N$, i.e. a nonconstant holomorphic map

$$\psi : S \longrightarrow \mathbb{G}(n-1, N),$$

where S is a compact Riemann surface. According to 1.5.1, this curve has a drill d with

$$1 \leq d \leq n \quad \text{and} \quad d \leq N - n + 1,$$

and there are derived curves (1.5.2)

$$\psi^{(1)} : S \longrightarrow \mathbb{G}(n-1+d, N) \quad \text{and} \quad \psi_{(1)} : S \longrightarrow \mathbb{G}(n-1-d, N).$$

These curves give rise to interesting varieties.

Proposition. *Let S be a compact Riemann surface and*

$$\psi : S \longrightarrow \mathbb{G}(n-1, N)$$

a curve of drill d with $\psi_{(1)}$ nonconstant if $d = 1$. We consider the following varieties:

$$X_{(1)} := \bigcup_{s \in S} \psi_{(1)}(s), \quad X := \bigcup_{s \in S} \psi(s), \quad X^{(1)} := \bigcup_{s \in S} \psi^{(1)}(s).$$

They have the following properties:

1) $X_{(1)} \subset X \subset X^{(1)} \subset \mathbb{P}_N$,

2) $\dim X = n$ *and* $\dim \gamma(X) = d$,

3) $X_{(1)} \subset \mathrm{Sing}\, X$ *and* $X_{(1)} = \mathrm{Foc}\, X$, *the focal variety of X (2.2.4),*

4) $X^{(1)} = \mathrm{Tan}\, X$, *the tangent variety of X (3.1.2).*

We call $X \subset \mathbb{P}_N$ a *twisted* $(n-1)$-*plane*.

In 1.5.5, we have seen that such curves ψ exist for given n and d, if $d \leq n$. The additional restriction for the codimension is

$$N - n \geq d - 1.$$

An example with $N = 4$, $n = 3$ and $d = 2$ is the twisted plane of 2.2.9. In this case

$$\psi = \lambda \oplus \beta^{(1)}, \quad \psi_{(1)} = \beta, \quad \psi^{(1)} = \lambda^{(1)} + \beta^{(2)}.$$

Proof. For almost every point $o \in S$ there is a coordinate neighborhood $U \subset S$ together with holomorphic maps

$$\varrho_i : U \longrightarrow \mathbb{C}^{N+1}, \quad \text{for } i = 0, \dots, n-1,$$

such that (1.5.2)

$(\varrho_0, \dots, \varrho_{n-1})$ is a local basis of ψ,

$(\varrho_d, \dots, \varrho_{n-1})$ is a local basis of $\psi_{(1)}$ and

$\left(\varrho_0, \dots, \varrho_{n-1}, \varrho_0', \dots, \varrho_{d-1}'\right)$ is a local basis of $\psi^{(1)}$.

This implies *1)*. By 2.2.2 the holomorphic map

$$\Phi : U \times \mathbb{C}^n \longrightarrow \widehat{X} \subset \mathbb{C}^{N+1}, \quad (s, t) \longmapsto \sum_{i=0}^{n-1} t_i \varrho_i(s),$$

is a parameterization for sufficiently small t. By definition of the drill, we know

$$T_{\Phi(s,t)} \widehat{X} = \text{span} \left\{ \varrho_0(s), \dots, \varrho_{n-1}(s), \sum_{i=0}^{d-1} t_i \varrho_i'(s) \right\} \subset \mathbb{C}^{N+1}. \qquad (*)$$

If we fix $s \in U$, then $(*)$ implies that the tangent spaces to smooth points of $\widehat{X}$ along $\psi(s)$ are precisely the $(n+1)$-dimensional vector spaces T such that

$$\psi(s) \subset T \subset \psi^{(1)}(s) \subset \mathbb{C}^{N+1}.$$

This implies *4)*. Since $\dim \psi(s) = n$ and $\dim \psi^{(1)}(s) = n + d$, this also implies for the Gauß map

$$\gamma : X \dashrightarrow \mathbb{G}(n, N)$$

of $X = \mathbb{P}(\widehat{X}) \subset \mathbb{P}_N$ that

$$\dim \gamma(\psi(s)) = d - 1,$$

if we consider $\psi(s)$ as a linear subspace of $X \subset \mathbb{P}_N$. Now *2)* follows from the assumption that ψ — and $\psi^{(1)}$ if $d = 1$ — are nonconstant.

Equation $(*)$ also implies that the tangent spaces to smooth points of $\widehat{X}$ are all the same along $(n - d + 1)$-dimensional vector spaces V such that

$$\psi_{(1)}(s) \subset V \subset \psi(s).$$

So the $(n - d)$-dimensional linear spaces $\mathbb{P}(V) \subset X \subset \mathbb{P}_N$ are contained in the fibers of the rational Gauß map, and since the general Gauß fiber is linear of the same dimension, they are equal in general.

Now we take a look at a point $p \in \psi_{(1)}(s) \subset X$ and its projective Zariski tangent space $\mathbb{T}_p X \subset \mathbb{P}_N$. For almost all s there are infinitely many different spaces $\mathbb{P}(V)$ passing p, and since they have different n-dimensional tangent spaces along different fibers $\mathbb{P}(V)$, we obtain

$$\dim \mathbb{T}_p X > n \, ;$$

hence, $p \in \operatorname{Sing} X$.

Finally, we compute the focal variety of X. We will use the map Φ to do so, even if it is not in the form described in 2.2.4, since $\Phi(s \times \mathbb{C}^n)$ is not a Gauß fiber of X. However, $\Phi(s \times \mathbb{C}^n) = \psi(s)$ is the union of the Gauß fibers V with $\psi_{(1)}(s) \subset V \subset \psi(s)$, and this will suffice for the computation. We note that for general $s \in S$ the map Φ is not immersive at (s, t) iff

$$\sum_{i=0}^{d-1} t_i \varrho_i'(s) \in \psi(s) \quad \Longleftrightarrow \quad \sum_{i=0}^{n-1} t_i \varrho_i'(s) \in \psi(s) \, .$$

By the definition of the focal hypersurface (2.2.4), this means that on a Gauß fiber $V = \Phi(s, \mathbb{C}(t_0, \dots, t_{d-1}) \times \mathbb{C}^{n-d})$ we have

$$\operatorname{Foc}(X, V) = \left\{ \sum_{i=0}^{n-1} t_i \varrho_i(s) : \sum_{i=0}^{n-1} t_i \varrho_i'(s) \in \psi(s) \right\} \cap V = \psi_{(1)}(s) \cap V = \psi_{(1)}(s) \, ;$$

hence, the focal variety of X is $X_{(1)}$.   □

The example of the twisted plane (2.2.9) shows that the inclusion $X_{(1)} \subset \operatorname{Sing} X$ may be proper.

The question under what condition a twisted $(n-1)$-plane is a cone is easy to answer (compare 1.5.5):

Remark. *The twisted $(n-1)$-plane*

$$X = \bigcup_{s \in S} \psi(s) \subset \mathbb{P}_N$$

is a cone iff $L := \bigcap_{s \in S} \psi(s) \subset \mathbb{P}_N$ is not empty.

More precisely, L is the vertex of the cone.

Proof. Assume E is the, possibly empty, vertex of the cone X (2.2.6). Then we obviously have $L \subset E$. Assume $E \not\subset \psi(s)$ for some $s \in S$. Since

$$E \# \psi(s) \subset X \quad \text{and} \quad \dim(E \# \psi(s)) \geq \dim \psi(s) + 1 = \dim X \, ,$$

we would obtain that $X = E \# \psi(s)$ is linear. Thus, finally $E = L$.   □

Together with the construction in 1.5.4, this finally yields the

Corollary. *Given integers* r, n, N *with*

$$0 \leq r \leq n \leq N \quad \text{and} \quad n + r \leq N + 1,$$

there exists a variety $X \subset \mathbb{P}_N$ *which is not a cone and*

$$\dim X = n, \quad \dim \gamma(X) = r.$$

Now one may ask under what condition the singularities of a twisted $(n-1)$-plane X are contained in a hyperplane $H \subset \mathbb{P}_N$. Then, in case X is not a cone, the affine part is smooth, developable, and not a cylinder.

By the above Remark the normal form (1.5.5) has no constant part; hence,

$$\psi = \varphi_1^{(a_1)} \oplus \ldots \varphi_d^{(a_d)}.$$

Since $X_{(1)} \subset \operatorname{Sing} X$, a necessary condition is that all curves φ_i with $a_i > 0$ have their image in H (use the Remark in 1.5.3).

2.4 The Second Fundamental Form

In 2.3.2, we have seen that in the generic case the Gauß rank is maximal, i.e. equal to the dimension of the variety. Now we will develop the basic tool to compute the Gauß rank and to describe the structure of varieties with not maximal Gauß rank.

2.4.1 Definition of the Second Fundamental Form

Basically, the second fundamental form is the differential of the Gauß map γ. However, in projective geometry there are two kinds of tangent spaces (1.1.1). If p is a smooth point of a variety $X \subset \mathbb{P}_N$, the Gauß image of p is the projective tangent space $\mathbb{T}_p X \subset \mathbb{P}_N$. The differential of γ is defined on the tangent vector space

$$T_p X \subset T_p \mathbb{P}_N \cong \mathbb{C}^N.$$

For this reason, we pass to the affine cone $\widehat{X} \subset \mathbb{C}^{N+1}$. More generally, we consider an arbitrary algebraic variety $Y \subset \mathbb{C}^{N+1}$ of dimension $n+1$. As in 2.3.1, we obtain a rational Gauß map

$$\gamma : Y \dashrightarrow \operatorname{Gr}(n+1, N+1) =: \mathbb{G}$$

with $\gamma(q) = T_q Y$ if $q \in Y$ is smooth. The differential of γ at q is a linear map

$$d_q \gamma : T_q Y \longrightarrow T_{T_q Y} \mathbb{G} \cong \operatorname{Hom}\left(T_q Y, \mathbb{C}^{N+1} / T_q Y\right).$$

If we use the canonical isomorphism of 1.4.2, the differential is described as follows. For $v, w \in T$ we choose the following gadgets:

– a holomorphic path $\alpha : S \to Y$ with $\alpha(o) = q$ and $\alpha'(o) = v$,

– a *vector field* $\varrho : S \to \mathbb{C}^{N+1}$ *along* α (i.e. $\varrho(s) \in T_{\alpha(s)}Y$ for every $s \in S$) with
   $\varrho(o) = w$.

If $\xi := \gamma \circ \alpha : S \to \mathbb{G}$, then $d_q\gamma(v) = \xi'(o)$, and since ϱ is a moving vector of ξ, 1.4.2
implies

$$d_q\gamma(v)(w) = \xi'(o)(w) = \varrho'(o) + T_qY \in \mathbb{C}^{N+1}/T_qY .$$

Restating the above construction, we have obtained a bilinear map

$$\mathbb{I}_q : T_qY \times T_qY \longrightarrow \mathbb{C}^{N+1}/T_qY \quad \text{with} \quad \mathbb{I}_q\left(\alpha'(o), \varrho(o)\right) = \varrho'(o) + T_qY$$

for every path α in Y with $\alpha(o) = q$ and every vector field ϱ along α. We call $\mathbb{I}_q$ the *second
fundamental form* of Y at q.

In particular, $\varrho = \alpha'$ is a vector field along α. This implies

$$\mathbb{I}_q\left(\alpha'(o), \alpha'(o)\right) = \alpha''(o) + T_qY .$$

By polarization, this quadratic map determines the bilinear map $\mathbb{I}_q$.

Recall that $\mathbb{I}_q$ is only defined if Y is smooth in q.

The image $N_qY := \mathbb{C}^{N+1}/T_qY$ of $\mathbb{I}_q$ is the *normal space* of Y at q in $\mathbb{C}^{N+1}$.

Proposition. $\mathbb{I}_q$ *is symmetric.*

Proof. We use a diagram

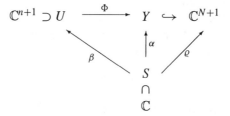

where U and S are neighborhoods of the origins o, Φ is a local parameterization with
$\Phi(o) = q$, α and β are holomorphic paths with $\alpha = \Phi \circ \beta$, and ϱ is a vector field along α.

For every $t = (t_0, \ldots , t_n) \in U$, we have a basis

$$\left(\frac{\partial\Phi}{\partial t_0}(t), \ldots , \frac{\partial\Phi}{\partial t_n}(t)\right) \quad \text{of } T_{\Phi(t)}Y \subset \mathbb{C}^{N+1} , \text{ and}$$

$$\alpha'(o) = \sum_{i=0}^{n} \beta_i'(o)\frac{\partial\Phi}{\partial t_i}(o) \in T_qY .$$

Furthermore, there are holomorphic functions $\varrho_0, \ldots, \varrho_n$ on S such that

$$\varrho(s) = \sum_{i=0}^{n} \varrho_i(s) \frac{\partial \Phi}{\partial t_i}(\beta(s)) .$$

Differentiating with respect to s and taking the value at $s = o$ yields

$$\varrho'(o) = \sum_{i=0}^{n} \varrho_i'(o) \frac{\partial \Phi}{\partial t_i}(o) + \sum_{i,j=0}^{n} \varrho_i(o) \frac{\partial^2 \Phi}{\partial t_i \partial t_j}(o) \beta_j'(o) .$$

Since the first sum is in $T_q Y$, we have

$$\mathbb{I}_q\left(\alpha'(o), \varrho(o)\right) = \sum_{i,j=0}^{n} \varrho_i(o) \frac{\partial^2 \Phi}{\partial t_i \partial t_j}(o) \beta_j'(o) + T_q Y = \mathbb{I}_q\left(\varrho(o), \alpha'(o)\right) . \qquad (*)$$

This also shows that the second fundamental form is described by the Hessian of Φ.   □

The curvature of a cone increases towards the origin; the second fundamental form behaves in a corresponding way:

Lemma. *Take a cone $Y \subset \mathbb{C}^{N+1}$ and a smooth point $q \in Y \backslash \{0\}$.*

a) *For every $\lambda \in \mathbb{C}^*$ we have*

$$T_{\lambda q} Y = T_q Y \quad and \ \mathbb{I}_{\lambda q} = \frac{1}{\lambda} \mathbb{I}_q .$$

b) *We have $q \in T_q Y$ and*

$$\mathbb{I}_q(q, T_q Y) = \mathbb{I}_q(T_q Y, q) = 0 .$$

Proof. a) Take a path α through q and a vector field ϱ along α as above. If we define

$$\tilde{\alpha}(s) := \lambda \alpha\left(\frac{s}{\lambda}\right) \quad and \quad \tilde{\varrho}(s) := \varrho\left(\frac{s}{\lambda}\right) ,$$

we obtain a path $\tilde{\alpha}$ through λq and a vector field $\tilde{\varrho}$ along $\tilde{\alpha}$ with

$$\tilde{\alpha}'(o) = \alpha'(o) \quad and \quad \tilde{\varrho}(o) = \varrho(o) ,$$

and the assertion follows from

$$\mathbb{I}_{\lambda q}\left(\tilde{\alpha}'(o), \tilde{\varrho}(o)\right) = \tilde{\varrho}'(o) + T_{\lambda q} Y = \frac{1}{\lambda} \varrho'(o) + T_q Y = \frac{1}{\lambda} \mathbb{I}_q(\alpha'(o), \varrho(o)) .$$

b) We use the path $\alpha(s) := q + sq$. For an arbitrary $v \in T_q Y$, the constant vector field $\varrho(s) := v$ is along α, since $T_{\alpha(s)} Y = T_q Y$; consequently,

$$\mathbb{I}_q(q, v) = \mathbb{I}_q\left(\alpha'(o), \varrho(o)\right) = \varrho'(o) + T_q Y = 0 + T_q Y .$$ □

2.4.2 The Degeneracy Space

The information we want to obtain here from the second fundamental form is just its degree of degeneracy. Thus, if $q \in Y \subset \mathbb{C}^{N+1}$ is a smooth point, we consider the *degeneracy space* (or *singular locus*, see 2.4.4)

$$K_q := \{v \in T_q Y : \mathbb{I}_q(v, w) = 0 \text{ for all } w \in T_q Y\}.$$

By the definition of $\mathbb{I}_q$, this is the kernel of the differential of the Gauß map

$$d_q \gamma : T_q Y \longrightarrow T_{\gamma(q)} \mathbb{G}.$$

The Gauß rank is determined by the dimension of a generic K_q; hence, we need an explicit description of these kernels.

Proposition. *Assume that the ideal of the variety $Y \subset \mathbb{C}^{N+1}$ is generated by polynomials $f_1, \ldots, f_m$ and that $q \in Y$ is a smooth point. Then*

$$K_q = \{v \in T_q Y : \text{Hess}_q f_i(v, w) = 0 \text{ for all } w \in T_q Y \text{ and } i = 1, \ldots, m\},$$

where $\text{Hess}_q f = \left(\frac{\partial^2 f}{\partial x_i \, \partial x_j}(q) \right)_{i,j}$ *denotes the Hesse matrix at q of the polynomial f.*

Proof. For $v, w \in T_q Y$ we choose a path $\alpha : S \to Y$ with $\alpha(o) = q$ and $\alpha'(o) = v$ and a vector field $\varrho : S \to \mathbb{C}^{N+1}$ along α with $\varrho(o) = w$. By 2.4.1, we know

$$\mathbb{I}_q(v, w) = \mathbb{I}_q(\alpha'(o), \varrho(o)) = \varrho'(o) + T_q Y,$$

and it is sufficient to show

$$\mathbb{I}_q(v, w) = 0 \iff \text{Hess}_q f_i(v, w) = 0 \quad \text{for } i = 1, \ldots, m. \tag{$*$}$$

By definition of the tangent space, we know

$$\text{grad}_{\alpha(s)} f_i(\varrho(s)) = 0 \quad \text{for } i = 1, \ldots, m$$

and sufficiently small s. Differentiating this relation and substituting $s = o$ yields

$$\text{Hess}_q f_i(\alpha'(o), \varrho(o)) + \text{grad}_q f_i(\varrho'(o)) = 0.$$

Now $\mathbb{I}_q(v, w) = 0$ means $\varrho'(o) \in T_q Y$, and this is equivalent to the vanishing of the brackets on the right for all i; this shows $(*)$. $\qquad \square$

This proposition will be used in 2.4.8 where we take a closer look at the singular points of the Gauß map.

In case of a hypersurface $Y \subset \mathbb{C}^{N+1}$, we need only one equation, i.e. $Y = V(f)$, and we assume f generates the ideal. Then the gradient

$$g_q := \left(\frac{\partial f}{\partial x_0}(q), \ldots, \frac{\partial f}{\partial x_N}(q) \right)$$

is nonzero in a smooth point $q \in Y$. Hence, the *extended Hesse matrix*

$$\overline{H}_q := \begin{pmatrix} 0 & g_q \\ {}^t g_q & \mathrm{Hess}_q\, f \end{pmatrix}$$

has rank at least 2. With exactly the same arguments as in 0.3, replacing $\mathbb{R}$ by $\mathbb{C}$ and using N instead of 2, we obtain the

Corollary. *If q is a general point of a hypersurface $Y \subset \mathbb{C}^{N+1}$, and $\gamma : Y \to \mathbb{P}_N^*$ is the Gauß map, then*

$$\mathrm{rank}\, d_q \gamma = \mathrm{rank}\, \overline{H}_q - 2 \,.$$

This is the easiest tool for computing the dimension of the Gauß image. It should be mentioned, that the zero in $\overline{H}_q$ can be replaced by f since $f(q) = 0$.

2.4.3 The Degeneracy Map

The lemma in 2.4.1 shows that the second fundamental form itself cannot be transported down directly from the affine cone to the projective variety, but the following immediate consequence of this lemma shows that a projective degeneracy space is easy to obtain.

Lemma. *If $Y \subset \mathbb{C}^{N+1}$ is a cone, $q \in Y \setminus \{0\}$ and $\lambda \in \mathbb{C}^*$, then*

$$q \in K_q = K_{\lambda q} \subset T_q Y \,,$$

where $T_q Y$ denotes the Zariski tangent space and $K_q \subset T_q Y$ is defined as in 2.4.2.

As a consequence, for a variety $X \subset \mathbb{P}_N$ and $p \in X$, we have a *degeneracy space* $\mathbb{K}_p$ such that

$$p \in \mathbb{K}_p \subset \mathbb{T}_p X \subset \mathbb{P}_N \,.$$

In case p is smooth, $\mathbb{T}_p X$ is the projective tangent space, and $\mathbb{K}_p$ is the projective degeneracy space of the second fundamental form.

By using the standard techniques for the construction of rational maps (1.1.2), we obtain the

Corollary. *Take a variety $X \subset \mathbb{P}_N$ of dimension n with Gauß map*

$$\gamma : X \dashrightarrow \mathbb{G}(n, N) \,.$$

If $r := \dim \gamma(X)$ and $k := n - r$, then there is a unique rational map

$$\kappa : X \dashrightarrow \mathbb{G}(k, N)$$

such that $\kappa(p) = \mathbb{K}_p$ for a general point $p \in X$.

We call κ the *degeneracy map* of X.

2.4.4 The Singular and Base Locus

For a variety $X \subset \mathbb{P}_N$, the second fundamental form was defined via the affine cone. For every smooth $q \in \widehat{X}$ lying over $p \in X$, we have a symmetric bilinear map

$$\widehat{\mathrm{II}}_q : T_q\widehat{X} \times T_q\widehat{X} \longrightarrow N_q\widehat{X} .$$

This gives a linear system

$$\widehat{\mathrm{II}}_q^* \subset \mathrm{Sym}^2(T_q\widehat{X})^*, \quad \dim \widehat{\mathrm{II}}_q^* \leq N - \dim X ,$$

which is the image of the dual map of $\widehat{\mathrm{II}}_q$

$$(N_q\widehat{X})^* = \left(\mathbb{C}^{N+1}/T_q\widehat{X}\right)^* \longrightarrow \mathrm{Sym}^2(T_q\widehat{X})^* .$$

Geometrically, every hyperplane $\widehat{H} \subset \mathbb{C}^{N+1}$ containing $T_q\widehat{X}$ determines a quadric cone $Q \subset T_q\widehat{X}$. Hence, we may write $Q \in \widehat{\mathrm{II}}_q^*$. The linear system has a *base locus*

$$\mathrm{Base}\,\widehat{\mathrm{II}}_q^* = \bigcap_{Q\in\widehat{\mathrm{II}}_q^*} Q = \{v \in T_q\widehat{X} : \widehat{\mathrm{II}}_q(v, v) = 0\} ,$$

which is a cone in $T_q\widehat{X}$, and a *singular locus*

$$\mathrm{Sing}\,\widehat{\mathrm{II}}_q^* = \bigcap_{Q\in\widehat{\mathrm{II}}_q^*} \mathrm{Sing}\, Q = K_q \subset T_q\widehat{X} ,$$

where K_q is the linear degeneracy space of $\widehat{\mathrm{II}}_q$ defined in 2.4.2. This is a geometric description of the second fundamental form on $\widehat{X}$, which does not make use of the fact that $\widehat{X}$ is a cone.

Further, since $\widehat{X}$ is a cone, $\widehat{\mathrm{II}}_q$ and $\widehat{\mathrm{II}}_{\lambda q}$ differ by a factor (2.4.1), but the linear systems are the same:

$$\widehat{\mathrm{II}}_q^* = \widehat{\mathrm{II}}_{\lambda q}^* \quad \text{for } q \in \widehat{X} \text{ and } \lambda \in \mathbb{C}^* .$$

Additionally, since $q \in K_q$, the quadratic forms induced by the linear system are invariant modulo $\mathbb{C}q$. Consequently, the linear system $\widehat{\mathrm{II}}_q^*$ determines a linear system

$$\mathrm{II}_p^* \subset \mathrm{Sym}^2(T_pX)^* ,$$

where $T_pX = T_q\widehat{X}/\mathbb{C}q$ is the tangent vector space of X at a smooth point $p \in X$. As above this means that every hyperplane H such that

$$T_pX \subset H \subset \mathbb{P}_N$$

determines a quadric cone $Q' \subset T_pX$ and, hence, a quadric $Q \subset \mathbb{P}(T_pX)$. Thus, finally, we have obtained the projective linear system

$$|\mathrm{II}_p^*| = \mathbb{P}(\mathrm{II}_p^*) \quad \text{of quadrics on } \mathbb{P}(T_pX) ,$$

determined by the second fundamental form. For its *singular locus* we have

$$\operatorname{Sing}|\mathbb{I}_p^*| = \bigcap_{Q \in \mathbb{I}_p^*} \operatorname{Sing} Q = \mathbb{P}(K_q/\mathbb{C}q) \subset \mathbb{P}(T_p X) .$$

The dimension of this linear subspace of $|\mathbb{I}_p^*|$ is determined by the Gauß map (2.5.2). The geometric meaning of the dimension of $\mathbb{I}_p^*$ itself is not at all obvious [GH2].

For our purpose, the linear system $\mathbb{I}_p^*$ obtained from a *bilinear* form contains sufficient geometric information. In some sense it shows how degenerate the "curvature" is. In order to measure the curvature continuously, one has to introduce the FUBINI-metric on $\mathbb{P}_N$; this metric is obtained by restricting the standard *hermitian* form on $\mathbb{C}^{N+1}$ to the sphere

$$S_{2N+1} = \{z \in \mathbb{C}^{N+1} : z \cdot \bar{z} = 1\}$$

and then transporting it down to $\mathbb{P}_N$.

2.4.5 The Codimension of a Uniruled Variety

There is an obvious relation between linear subspaces $L \subset X$ and the base loci along L.

Remark. *If $X \subset \mathbb{P}_N$ is a variety and $L \subset X$ is linear, then for every $p \in L$, which is smooth in X, we have $L \subset \mathbb{B}_p := \mathbb{P}(\operatorname{Base} \widehat{\mathbb{I}}_q^*)$.*

Proof. Take $q \in \widehat{L} \subset \widehat{X}$ over p, $v \in \widehat{L}$ arbitrary and the path

$$\alpha : \mathbb{C} \longrightarrow \widehat{X} , \qquad s \longmapsto q + sv .$$

Then $\alpha(o) = q$ and $\alpha'(o) = v$, hence

$$\widehat{\mathbb{I}}_q(v, v) = \widehat{\mathbb{I}}_q\left(\alpha'(o), \alpha'(o)\right) = \alpha''(o) + T_q\widehat{X} = 0 + T_q\widehat{X} . \qquad \square$$

This can be used to obtain an estimate for the codimension of a uniruled variety (see [R] and [L₁]).

Proposition. *Assume $X \subset \mathbb{P}_N$ is a variety of dimension n and of Gauß defect $d = \delta_\gamma(X)$, with a uniruling by $(k + d)$-planes containing the Gauß uniruling. Then*

$$\operatorname{codim} X \geq \frac{k}{n - d - k} .$$

In particular, if X is smooth and uniruled by k-planes, then

$$\operatorname{codim} X \geq \frac{k}{n - k} .$$

Recall that $n - d - k$ is the dimension of the base of the uniruling.

Proof. If X is smooth, then $d = 0$ by the Theorem of ZAK (2.3.2). Hence, it is sufficient to prove the first assertion.

For every smooth $p \in X$ we have a linear space L of dimension $k+d$ such that $p \in L \subset X$. By the above remark

$$\mathbb{K}_p \subset L \subset \mathbb{B}_p .$$

We need the following lemma.

Lemma. *Let $E \subset \mathrm{Sym}^2 \left(\mathbb{C}^{m+l} \right)^*$ be a linear system of quadrics and denote by*

$$S \subset B \subset \mathbb{C}^{m+l}$$

its singular locus and base locus, where $\dim S = l$. *If there is a $(k + l)$-dimensional linear space L such that $S \subset L \subset B$, and if $\dim E > 0$, then*

$$(m - k) \dim E \geq k .$$

Proof of the Lemma. By passing to the quotient $\mathbb{C}^{m+l}/S$, we may assume $l = 0$. Then we may assume that $e_1, \dots , e_k$ form a base of $L \subset B$. Consequently, every quadric $Q \in E$ can be given by a matrix of the form

$$A^* = \left(\begin{array}{c|c} 0 & A \\ \hline {}^t\!A & * \end{array} \right) ,$$

where A is a $k \times (m - k)$-matrix. Since the case $m - k \geq k$ is trivial, we may assume $m < 2k$.

The singular locus $\mathrm{Sing}\, Q$ is the degeneracy space of the matrix A^*, and

$$L \cap \mathrm{Sing}\, Q = \mathrm{Ker} \left(\begin{array}{c} 0 \\ {}^t\!A \end{array} \right) ; \quad \text{hence,}$$

$$\dim(L \cap \mathrm{Sing}\, Q) = k - \mathrm{rank}\, A \geq k - (m - k) .$$

For the entire linear system this implies

$$0 = \dim \left(L \cap \bigcap_{Q \in E} \mathrm{Sing}\, Q \right) \geq k - (m - k) \dim E . \qquad \square$$

If we put $m = n - d$, $E = \widehat{\mathbb{I}}_q^*$ and use codim $X \geq \dim \widehat{\mathbb{I}}_q^*$, the inequality of the proposition follows. $\qquad \square$

2.4.6 Fibers of the Gauß Map

For every smooth point p of a variety $X \subset \mathbb{P}_N$, we have two linear spaces

- the degeneracy space $\mathbb{K}_p$ of the second fundamental form, which is the kernel of the differential of the Gauß map γ (2.4.3)
- the fiber $\gamma^{-1}(\gamma(p))$ of the Gauß map, which is linear for almost all p by 2.3.2.

Now we prove the basic fact that for almost all points p these two linear spaces are the same.

Theorem. *If $X \subset \mathbb{P}_N$ is a variety with Gauß map*

$$\gamma : X \dashrightarrow \mathbb{G}(n, N) ,$$

then, for almost all $p \in X$, the fiber $\gamma^{-1}(\gamma(p))$ and the degeneracy space $\mathbb{K}_p \subset \mathbb{T}_p X$ are equal.

In particular, the fibers of γ can be described in terms of gradients and Hessians of the polynomials defining X (2.4.2).

The *proof* is easy if we use the Linearity Theorem of 2.3.2, proved by methods of global projective geometry. For a generic point $p \in X$ we have

- the fiber $F_p := \gamma^{-1}(\gamma(p))$ and

- the degeneracy space $\mathbb{K}_p \subset \mathbb{T}_p X$

Now, in general, if the fiber F_p is smooth in p, then

$$\mathbb{T}_p F_p \subset \mathbb{K}_p$$

by definition of the differential. In our case F_p and $\mathbb{K}_p$ are linear subspaces of the same dimension of $\mathbb{T}_p X$, hence $F_p = \mathbb{K}_p$.

We can give an *alternative proof*, which includes a new proof of the linearity of the Gauß fibers by generalizing the arguments of 0.2. However, it is purely local and does not show the connectedness of the Gauß fibers. More precisely, it is sufficient to prove the following

Lemma. *If $Y \subset \mathbb{C}^{N+1}$ is a variety, $q \in Y$ is a general point, and γ is regular in q, then the affine space $q + K_q$ is an irreducible component of $\gamma^{-1}(\gamma(q))$.*

Proof. If $\dim Y = n + 1$, $\dim \gamma(Y) = r$, and $k = n - r$, then by assumption

$$\operatorname{rank} d_y \gamma = r$$

for all y in a neighborhood W of q in Y. By the implicit function theorem, there is a holomorphic parameterization

$$\Phi : U \longrightarrow W \subset Y, \quad t = (t_0, \ldots, t_k, t_{k+1}, \ldots, t_n) \longmapsto \Phi(t),$$

with $o \in U \subset \mathbb{C}^{n+1}$ and $\Phi(o) = q$ such that $(t_0, \ldots, t_k)$ parameterizes a fiber of γ in W for fixed $(t_{k+1}, \ldots, t_n)$. In particular, for every $t \in U$

$$\left(\frac{\partial \Phi}{\partial t_0}(t), \ldots, \frac{\partial \Phi}{\partial t_k}(t) \right) \quad \text{is a basis of } K_{\Phi(t)}, \quad \text{and}$$

$$\left(\frac{\partial \Phi}{\partial t_0}(t), \ldots, \frac{\partial \Phi}{\partial t_n}(t) \right) \quad \text{is a basis of } T_{\Phi(t)}Y.$$

Define $U_1 := \{ t \in U : t_{k+1} = \ldots = t_n = 0 \}$ and $\varphi := \Phi | U_1$. If we can prove

$$\frac{\partial^l \varphi}{\partial t_{i_1} \ldots \partial t_{i_l}}(t) \in K_{\varphi(t)} \quad \text{for } i_1, \ldots, i_l \in \{0, \ldots, k\} \tag{K_l}$$

for every $l \geq 1$ and for every $t \in U$, then by looking at the Taylor series of φ at o this implies

$$\varphi(U_1) \subset q + K_{\varphi(0)}.$$

Since the dimension of $\gamma^{-1}(\gamma(q))$ at q is equal to the dimension of K_q, the assertion follows.

In order to prove (K_l), we use the auxiliary statement

$$\frac{\partial^l \varphi}{\partial t_{i_1} \ldots \partial t_{i_{l-1}} \partial t_j}(t) \in T_{\varphi(t)}Y \quad \text{for } i_1, \ldots, i_{l-1} \in \{0, \ldots, k\} \text{ and } j \in \{0, \ldots, n\} \tag{T_l}$$

for $l \geq 1$, and we prove (K_l) and (T_l) by induction. The case $l = 1$ is obvious by construction of φ. We prove

$$(T_l) \Rightarrow (T_{l+1}) \Rightarrow (K_l).$$

First, (T_l) implies

$$\text{grad}_{\varphi(t)} f \left(\frac{\partial^l \varphi}{\partial t_{i_1} \ldots \partial t_{i_{l-1}} \partial t_j}(t) \right) = 0$$

for the indices as above and for every polynomial f in the ideal of Y. Differentiating this relation with respect to t_i yields

$$\begin{aligned} (1) \quad &:= \quad \text{Hess}_{\varphi(t)} f \left(\frac{\partial \varphi}{\partial t_i}(t), \frac{\partial^l \varphi}{\partial t_{i_1} \ldots \partial t_{i_{l-1}} \partial t_j}(t) \right) \\ &= \quad -\text{grad}_{\varphi(t)} f \left(\frac{\partial^{l+1} \varphi}{\partial t_{i_1} \ldots \partial t_{i_{l-1}} \partial t_j \partial t_i}(t) \right) =: (2). \end{aligned}$$

If $i \in \{0, \ldots, k\}$ and $j \in \{0, \ldots, n\}$, then (K_1) and the definition of $K_{\varphi(t)}$ implies $(1) = 0$; hence, $(2) = 0$, implying (T_{l+1}).

If $i \in \{0, \dots, n\}$ and $j \in \{0, \dots, k\}$, then (T_{l+1}) says $(2) = 0$; hence, $(1) = 0$, which means (K_l). $\qquad\square$

Now we compare the Gauß map γ and the degeneracy map (2.4.3)

$$\kappa : X \dashrightarrow \mathbb{G}(k, N), \quad p \longmapsto \mathbb{K}_p .$$

If $\dim X = n$ and $\dim \gamma(X) = r$, then $n = k + r$. Since γ and κ have the same fibers almost everywhere, we have $\dim \kappa(X) = \dim \gamma(X)$, and there is a birational *shrinking map* σ, such that

$$
\begin{array}{ccc}
& & \gamma(X) \subset \mathbb{G}(n, N) \\
& {\scriptstyle \gamma} \nearrow & \quad | \\
X & & \quad | \; \sigma \\
& {\scriptstyle \kappa} \searrow & \quad \downarrow \\
& & \kappa(X) \subset \mathbb{G}(k, N)
\end{array}
$$

commutes. More explicitely, for a general point $p \in X$, we have

$$\sigma(\mathbb{T}_p X) = \mathbb{K}_p \subset \mathbb{T}_p X .$$

The variety $\kappa(X)$ is the base of the almost ruling determined by the fibers of γ; it is called the *Gauß almost ruling* of X.

2.4.7 Characterization of Gauß Images

In the previous section, we compared the images of X via the Gauß map and the degeneracy map. Since $\kappa(X)$ is a base of the Gauß almost ruling,

$$X = \bigcup_{K \in \kappa(X)} K .$$

It is quite natural to ask how X can be reconstructed from the Gauß image $\gamma(X)$, or, more generally, is any given variety $Y \subset \mathbb{G}(n, N)$ the Gauß image of some $X \subset \mathbb{P}_N$, and how can X be reconstructed from Y. The key to the solution is the observation that the shrinking map σ, from paragraph 2.4.6, can be described without using X.

If $Y \subset \mathbb{G}(n, N)$ is an arbitrary variety, then a point $E \in Y$ corresponds to linear spaces $E \subset \mathbb{P}_N$ and $\widehat{E} \subset \mathbb{C}^{N+1}$. According to paragraph 1.4.2, we have a Zariski tangent space

$$T_E Y \subset T_E \mathbb{G}(n, N) \cong \mathrm{Hom}\,(\widehat{E}, \mathbb{C}^{N+1}/\widehat{E}) ,$$

so we can define a linear subspace

$$\sigma(E) := \mathbb{P}\left(\bigcap_{f \in T_E Y} \mathrm{Ker}\, f \right) \subset E .$$

If k denotes the generic dimension of $\sigma(E)$, we obtain a rational map

$$\sigma : Y \dashrightarrow \mathbb{G}(k, N) ,$$

which we call *shrinking map* of Y. Now there is an obvious candidate for $X \subset \mathbb{P}_N$, the only obstacles are the dimensions of the spaces occurring in the construction.

Theorem. *Take a variety $Y \subset \mathbb{G}(n, N)$ with shrinking map $\sigma : Y \dashrightarrow \mathbb{G}(k, N)$ and define*

$$X := \bigcup_{K \in \sigma(Y)} K \subset \mathbb{P}_N.$$

Then the following conditions are equivalent:

   i) *Y is the image of a Gauß map,*

  ii) *Y is the image of the Gauß map of X,*

 iii) *$k + \dim Y = n$ and $\dim X = n$.*

In condition *iii)*, $\dim X = n$ means only that X has the dimension expected from the naive dimension count $k + \dim Y = n$. It can very often be verified by a first order calculation, see [H, 16.13]. Without this additional condition the Theorem is wrong; for the easiest example, consider a curve $Y \subset \mathbb{G}(1, N)$ of lines such that all lines pass through a common point $p \in \mathbb{P}_N$. Then Y is certainly not the image of a Gauß map, but σ is the constant map given by p, and $k + \dim Y = n$ holds.

Proof of the Theorem. i) $\Rightarrow$ ii),iii) Let Y be the image of the Gauß map of the variety Z. Consider the affine Gauß map

$$\gamma_{\widehat{Z}} : \widehat{Z} \dashrightarrow \mathrm{Gr}(n + 1, N + 1)$$

and a general $q \in \widehat{Z}$. For $E = \gamma_{\widehat{Z}}(q)$, the equation

$$\kappa_Z(q) = \mathrm{Sing}\,\widehat{\mathrm{I\!I}}_{\widehat{Z},q} = \bigcap_{f \in \mathrm{Im}\, d\gamma_{\widehat{Z},q}} \mathrm{Ker}\, f = \bigcap_{f \in T_E Y} \mathrm{Ker}\, f = \sigma(E)$$

is valid. Hence,

$$Z = \bigcup_{K \in \kappa_Z(Z)} K = \bigcup_{K \in \sigma(Y)} K = X.$$

In particular, $\dim X = n$ and $k = n - \dim Y$, since $\kappa_Z(q)$ is a general fiber of the Gauß map.

iii) $\Rightarrow$ ii) Let $\varphi : S \to Y$ be a local parameterization of Y near $\varphi(o) = E$ where $S \subset \mathbb{C}^{n-k}$. After shrinking S, there exist $\varrho_0, \dots, \varrho_k : S \to \mathbb{C}^{N+1}$ such that

$$\sigma \circ \varphi(s) = \mathrm{span}\,\{\varrho_0(s), \dots, \varrho_k(s)\} \subset \varphi(s) \subset \mathbb{P}_N.$$

For $\frac{\partial \varphi}{\partial s_j}(s) \in T_{\varphi(s)}Y \subset \mathrm{Hom}\,(\varphi(s), \mathbb{C}^{N+1}/\varphi(s))$, we have by the definition of σ

$$\frac{\partial \varphi}{\partial s_j}(s)(\varrho_i(s)) = 0 + \varphi(s)\,.$$

By the isomorphism $T_{\varphi(s)}\mathbb{G}(n, N) \cong \mathrm{Hom}\,(\varphi(s), \mathbb{C}^{N+1}/\varphi(s))$

$$\frac{\partial \varphi}{\partial s_j}(s)(\varrho_i(s)) = \frac{\partial \varrho_i}{\partial s_j}(s) + \varphi(s) = 0 + \varphi(s) \quad \Longrightarrow \quad \frac{\partial \varrho_i}{\partial s_j}(s) \in \varphi(s)\,. \qquad (*)$$

By definition, $\widehat{X}$ is locally the image of

$$\psi : S \times \mathbb{C}^{k+1} \longrightarrow \widehat{X}, \quad (s, t) \longmapsto \sum_{i=0}^{k} t_i \varrho_i(s)\,.$$

Hence, the tangent space of $\widehat{X}$ at a smooth point $\psi(s, t)$ is given by

$$\mathbb{T}_{\psi(s,t)}\widehat{X} = \mathrm{span}\left\{\varrho_0(s), \ldots, \varrho_k(s), \sum_{i=0}^{k} t_i \frac{\partial \varrho_i}{\partial s_1}(s), \ldots, \sum_{i=0}^{k} t_i \frac{\partial \varrho_i}{\partial s_{n-k}}(s)\right\}\,.$$

By $(*)$ above, $\gamma_{\widehat{X}}(\psi(s, t)) = \mathbb{T}_{\psi(s,t)}\widehat{X} \subset \varphi(s)$. Since both spaces have the same dimension by $\dim X = n$, we get $\gamma_{\widehat{X}}(\psi(s, t)) = \varphi(s)$. Therefore $\mathrm{Im}\,\gamma_X = Y$. $\qquad \square$

For the special case when Y is a curve, we have additional equivalent conditions:

Corollary. *For a curve $C \subset \mathbb{G}(n, N)$, the following conditions are equivalent:*

i) *C is the image of a Gauß map,*

ii) *a parameterization $\psi : S \to \mathbb{G}(n, N)$ of C has the normal form $\psi = \varphi^{(a)} \oplus L$ with $a \geq 1$ (1.5.5),*

iii) *for all smooth points $E \in C$*

$$\mathbb{T}_E C \subset \mathbb{G}(n, N) \subset \mathbb{P}(\textstyle\bigwedge^{n+1}\mathbb{C}^{N+1})$$

   holds and

$$\dim \bigcap_{E \in C} E \leq n - 2\,.$$

If the conditions are fulfilled, then C is the Gauß image of

$$X = \bigcup_{s \in S}\left(\varphi^{(a-1)}(s) \oplus L\right)\,.$$

Proof. ii) is equivalent to *iii)* by Theorem 2.2.10 and the obvious fact that $a \geq 1$ is equivalent to

$$L = \bigcap_{E \in C} E \leq n - 2 .$$

If $\psi : S \to \mathbb{G}(n, N)$ is a parameterization of the curve C, then for a general $s \in S$

$$\sigma(\psi(s)) = \operatorname{Ker} f \quad \text{for } f \in T_{\psi(s)} C \setminus \{0\} .$$

By the isomorphism $T_{\psi(s)} \mathbb{G}(n, N) \cong \operatorname{Hom}(\psi(s), \mathbb{C}^{N+1} / \psi(s))$, we have

$$\sigma(\psi(s)) = \operatorname{Ker} f = \{\varrho(s) : \text{ for some } \varrho \in \psi \text{ with } \varrho'(s) \in \psi(s)\} = \psi_{(1)}(s) .$$

The theorem tells us that C is a Gauß image if and only if both of the following conditions hold:

a) $\sigma \circ \psi = \psi_{(1)}$ is a map to $\mathbb{G}(n - 1, N)$, i.e. the drill of ψ is 1, so ψ has normal form $\varphi^{(a)} \oplus L$;

b) the dimension of

$$\bigcup_{s \in S} \sigma \circ \psi(s)$$

is n, i.e. $\sigma \circ \psi = \psi_{(1)}$ is not constant. This is equivalent to $a \geq 1$. (1.5.5)

Finally, the additional statement follows from $\sigma \circ \psi = \psi_{(1)} = \varphi^{(a-1)} \oplus L$. $\qquad \square$

2.4.8 Singularities of the Gauß Map

The Gauß map γ of a variety $X \subset \mathbb{P}_N$ is regular on the smooth part X_{sm} and is extended to the singular locus Sing X as a rational map by the usual closing procedure. In general, γ may have singular points on X_{sm}; for instance, flexes in case of curves or planar points in case of surfaces. The behavior of γ on Sing X is much more difficult to understand. One thing is relatively easy to clarify: the position of the (linear) fibers of γ relative to the singularities of γ and X. First, we have to specify what we mean by a "singular point" of the rational map γ.

For this purpose we consider the affine cone $\widehat{X} \subset \mathbb{C}^{N+1}$ with projection

$$\pi : \widehat{X} \setminus \{0\} \longrightarrow X, \quad q \longmapsto \mathbb{C}q ,$$

and the algebraic space

$$T\widehat{X} \subset \widehat{X} \times \mathbb{C}^{N+1} \quad \text{with projection } \tau : T\widehat{X} \longrightarrow \widehat{X} .$$

For every $q \in \widehat{X}$, we have $\tau^{-1}(q) = T_q \widehat{X} \subset \{q\} \times \mathbb{C}^{N+1}$, the Zariski tangent space. Since it is defined by the gradients of the homogeneous polynomials defining $\widehat{X}$, the set

$$S := \{q \in \widehat{X} : \dim T_q \widehat{X} > \dim \widehat{X}\}$$

is an algebraic cone (the rank condition can be expressed by determinants); hence,

$$\text{Sing } X = \pi(S \setminus \{0\}) \subset X$$

is an algebraic set.

Now we consider the Gauß map

$$\gamma : X \dashrightarrow \mathbb{G}(k, N)$$

and the degeneracy space $K_q \subset T_q \widehat{X}$, which can be described via Hesse matrices (2.4.2). Hence, the set

$$S := \text{closure of } \{q \in \widehat{X} \setminus S : \dim K_q > k + 1\}$$

is an algebraic cone. Defining

$$\text{Sing } \gamma := \pi(S \setminus \{0\}) \subset X,$$

we obtain an algebraic set, such that

$$\text{Sing } \gamma \cap X_{\text{sm}} = \{p \in X_{\text{sm}} : \text{rank } d_p \gamma < \dim \gamma(X)\}.$$

We call $\text{Sing } \gamma \subset X$ the *singular locus of* γ. The promised result is:

Proposition. *Let* $X \subset \mathbb{P}_N$ *be a variety with Gauß map*

$$\gamma : X \dashrightarrow \mathbb{G}(n, N)$$

and degeneracy map

$$\kappa : X \dashrightarrow \mathbb{G}(k, N).$$

Then for every leaf E of the Gauß ruling, i.e. for every linear space $E \in \kappa(X)$, we have either

$$E \subset \text{Sing } \gamma \quad \text{or} \quad E \cap \text{Sing } \gamma \subset \text{Sing } X.$$

In particular,

$$\text{Sing } \gamma = \bigcup_{E \in B} E, \quad \text{where } B := \{E \in \kappa(X) : E \subset \text{Sing } \gamma\},$$

and $B \subset \mathbb{G}(k, N)$ is algebraic.

Proof. It suffices to show the following:

if $p \in E \cap (X_{\text{sm}} \setminus \text{Sing } \gamma)$ and $x \in E \cap X_{\text{sm}}$, then $x \notin \text{Sing } \gamma$.

For this purpose, we use the following

– the affine cone $\widehat{X} \subset \mathbb{C}^{N+1}$ with points q and y lying over p and x,

– an open neighborhood S of the origin o in $\mathbb{C}^r$, where $r = \dim \gamma(X) = \dim \kappa(X)$ and

– an adapted parameterization (2.2.2)

$$\Phi : S \times \mathbb{C}^{k+1} \longrightarrow \widehat{X}, \quad (s, t) \longmapsto \sum_{i=0}^{k} t_i \varrho_i(s),$$

such that $\Phi(o, e_0) = q$ for $e_0 = (1, 0, \ldots, 0)$, which is an immersion for all $s \in S$ and for $t \in \mathbb{C}^{k+1}$ sufficiently close to e_0.

Since $y \in \widehat{E} = \Phi(o, \mathbb{C}^{k+1})$, there is a $o \neq \xi \in \mathbb{C}^{k+1}$ such that $y = \Phi(o, \xi)$, but Φ need not be immersive at (o, ξ). Nevertheless, it is possible to control

$$K_y \subset T_y \widehat{X} \quad \text{by} \quad K_q \subset T_q \widehat{X} = T_y \widehat{X}.$$

More precisely, we show $K_y = K_q$. By our assumption, we know $\dim K_y \geq \dim K_q$. Hence, it suffices to show $K_y \subset K_q$.

If I denotes the ideal of $\widehat{X} \subset \mathbb{C}^{N+1}$, then for every smooth $z \in \widehat{E}$ we know (2.4.2)

$$K_z = \{v \in T_q \widehat{X} : \operatorname{Hess}_z f(v, w) = 0 \quad \text{for all } w \in T_q \widehat{X} \text{ and all } f \in I\},$$

and we use the basis (2.2.4)

$$\left(\frac{\partial \varrho_0}{\partial s_1}(o), \ldots, \frac{\partial \varrho_0}{\partial s_r}(o), \varrho_0(o), \ldots, \varrho_k(o) \right) \quad \text{of } T_q \widehat{X}.$$

Consequently, for every $v \in K_y$, we have to show

$$\operatorname{Hess}_q f\left(v, \varrho_j(o)\right) = 0 \quad \text{for } j = 0, \ldots, k \quad \text{and} \tag{1}$$

$$\operatorname{Hess}_q f\left(v, \frac{\partial \varrho_0}{\partial s_i}(o)\right) = 0 \quad \text{for } i = 0, \ldots, r. \tag{2}$$

Equation (1) is obvious since $v \in K_y \subset T_q \widehat{X}$ and $\varrho_j(o) \in K_q$. In order to prove (2), we use a vector field

$$\eta : S \longrightarrow \mathbb{C}^{N+1}, \quad \text{such that } \eta(o) = v \text{ and } \eta(s) \in T_{\Phi(s, e_0)} \widehat{X}.$$

If $\widehat{X}$ is smooth in $\Phi(s, t)$, then

$$T_{\Phi(s, t)} \widehat{X} = T_{\Phi(s, e_0)} \widehat{X}.$$

Hence, for every $s \in S, t \in \mathbb{C}^{k+1}$, and $f \in I$ we have

$$\operatorname{grad}_{\Phi(s, t)} f(\eta(s)) = 0.$$

Differentiating this relation with respect to s_i yields

$$\mathrm{Hess}_{\Phi(s,t)} f\left(\frac{\partial \Phi}{\partial s_i}(s,t), \eta(s)\right) + \mathrm{grad}_{\Phi(s,t)} f\left(\frac{\partial \eta}{\partial s_i}(s)\right) = 0. \tag{3}$$

This relation can be written shorter as

$$h(s,t) + g(s,t) = 0, \tag{3'}$$

where g and h are holomorphic functions on $S \times \mathbb{C}^{k+1}$. With this notation, we finally have to show that $v \in K_y$ implies $h(o, e_0) = 0$ since

$$\frac{\partial \Phi}{\partial s_i}(o, e_0) = \frac{\partial \varrho_0}{\partial s_i}(o).$$

Now $v \in K_y$ implies $h(o, \xi) = 0$, and $g(o, \xi) = 0$ by (3'). This yields

$$\frac{\partial \eta}{\partial s_i}(o) \in T_y \widehat{X} = T_q \widehat{X};$$

hence, $g(o, e_0) = 0$, and again by (3'), we obtain $h(o, e_0) = 0$.

We still have to show that $\mathrm{Sing}\, \gamma$ is a union of leaves. Since, according to (2.1.2),

$$B = \kappa(X) \cap F_k(\mathrm{Sing}\, \gamma), \quad \text{we know} \quad \bigcup_{E \in B} E \subset \mathrm{Sing}\, \gamma \subset X$$

is algebraic. Hence, it is sufficient to show

$$\mathrm{Sing}\, \gamma \setminus \mathrm{Sing}\, X \subset \bigcup_{E \in B} E.$$

Every point $x \in X$ is contained in some leaf $E \in \kappa(X)$; if $x \in \mathrm{Sing}\, \gamma \setminus \mathrm{Sing}\, X$, then the only possibility is $E \subset \mathrm{Sing}\, \gamma$. This proves the assertion.   □

It is not at all clear if the almost ruling of X defined by the Gauß fibers induces a uniruling of $\mathrm{Sing}\, \gamma$, i.e. if $\mathrm{Sing}\, \gamma$ has the expected dimension $\dim B + k$.

2.5 Gauß Defect and Dual Defect

In the paragraphs 2.3.4 and 2.1.4, we had defined two measures of special behavior of a variety X, the Gauß defect $\delta_\gamma(X)$ and the dual defect $\delta_*(X)$. It was easy to see (2.3.4) that

$$0 \le \delta_\gamma(X) \le \delta_*(X) \le \dim X,$$

and that only linear varieties have maximal defects.

In this section, we want to give a more precise information on these two numbers. As a consequence, we will be able to classify developable hypersurfaces, where the classification is especially explicit for Gauß rank at most two (2.5.6).

2.5.1 Dual Defect of Segre Varieties

We first take a look at the most important examples of ruled varieties, the *Segre varieties*, i.e. the images of the *Segre map*

$$\sigma : \quad \mathbb{P}_m \times \mathbb{P}_n \quad \longrightarrow \quad \mathbb{P}_{(m+1)(n+1)-1} ,$$
$$(x_0 : \ldots : x_m; y_0 : \ldots : y_n) \longmapsto (\ldots : x_i y_j : \ldots) .$$

We write $X_{m,n} := \sigma(\mathbb{P}_m \times \mathbb{P}_n)$ and $N := (m+1)(n+1) - 1$.

The Segre varieties $X_{m,n} \subset \mathbb{P}_N$ are all smooth; hence,

$$\delta_\gamma(X_{m,n}) = 0$$

for all m, n. This is an immediate consequence of ZAK's theorem (2.3.3), but can also be verified directly. For the dual defect, we have the

Proposition. *If we assume $m \le n$, then*

$$\delta_*(X_{m,n}) = n - m .$$

In particular, $X_{m,n}$ is ordinary iff $m = n$.

Proof. We consider the vector space $M = \mathbb{C}^{(m+1)(n+1)}$ as the space of $(m+1) \times (n+1)$–matrices $Z = (z_{ij})$. The affine Segre cone

$$\widehat{X}_{m,n} \subset M$$

is the variety of matrices of rank ≤ 1, since points of the Segre cone correspond to decomposable tensors.

For the dual projective space, we have $\mathbb{P}_N^* = \mathbb{P}(M^*)$, where M^* can again be considered as the space of $(m+1) \times (n+1)$–matrices $A = (a_{ij})$ and the linear form on M is defined by

$$A(Z) = \sum_{i,j} a_{ij} z_{ij} .$$

Now it is sufficient to prove

$$\widehat{X}_{m,n}^* = \{A \in M^* : \operatorname{rank} A \le m\} , \tag{$*$}$$

since then a simple counting argument shows

$$\dim \widehat{X}_{m,n}^* = m(n+2) .$$

Since $X_{m,n}$ and $X_{m,n}^*$ are homogeneous [H, 15.22], the assertion $(*)$ is proved if we can find matrices $A \in \widehat{X}_{m,n}^*$ of rank m. For this purpose, we fix the point

$$Z_0 \in \widehat{X}_{m,n} \quad \text{with } z_{00} = 1 \quad \text{and } z_{ij} = 0 \quad \text{otherwise} .$$

Now the tangent space T of $\widehat{X}_{m,n}$ at Z_0 is spanned by matrices of the form

$$\begin{pmatrix} \lambda_0 & \\ \vdots & \text{\huge 0} \\ \lambda_m & \end{pmatrix} \quad \text{and} \quad \begin{pmatrix} \mu_0 \dots \mu_n \\ \text{\huge 0} \end{pmatrix}$$

with $\lambda_i, \mu_j \in \mathbb{C}$. Hence, $A(T) = 0$ is equivalent to

$$a_{ij} = 0 \quad \text{if } i = 0 \text{ or } j = 0.$$

This implies rank $A \le m$.   □

In the case of $m = n$ this proof also shows

$$\widehat{X}^*_{m,n} = \{Z \in \mathbb{P}^*_N : \det Z = 0\}.$$

With the same arguments as above the reader may also compute the dual defect of

$$X = \mathbb{P}\{Z \in M : \text{rank } Z \le k\}$$

for $0 \le k \le m \le n$.

2.5.2 Gauß Defect and Singular Locus

The second fundamental form can be interpreted as a linear system of quadrics (2.4.4). From this point of view, the Gauß defect δ_γ of a variety X is caused by the singularities of all these quadrics. The same is true for the dual defect, but in a more subtle way (2.5.3).

Since $\mathbb{K}_p$ is generically the fiber of the Gauß map, we know

$$\delta_\gamma(X) = \min\{\dim \mathbb{K}_p : p \in X\}.$$

On the other hand, the second fundamental form determines a linear system $|\mathbb{I}^*_p|$ of quadrics

$$Q \subset \mathbb{P}(T_p X) \quad \text{with Sing} \, |\mathbb{I}^*_p| = \mathbb{P}(K_q/\mathbb{C}q)$$

for every smooth point $p \in X$ (2.4.4). With the convention

$$\text{Sing} \, |\mathbb{I}^*_p| = \mathbb{P}(T_p X) \quad \text{for } |\mathbb{I}^*_p| = \emptyset,$$

this immediately implies the following proposition.

Proposition. *For almost every point p of a variety $X \subset \mathbb{P}_N$, we have*

$$\delta_\gamma(X) = \dim(\text{Sing} \, |\mathbb{I}^*_p|) + 1.$$

In particular, $\delta_\gamma(X) = 0$ iff there is a smooth $p \in X$ such that $\text{Sing} \, |\mathbb{I}^*_p| = \emptyset$.

2.5.3 Dual Defect and Singular Locus

Now we relate the dual defect to the singularities of the quadrics in the linear system associated to the second fundamental form. While the Gauß defect is related to the singular locus of the whole system, the dual defect depends on the singular loci of the individual members of the system. This explains why the dual defect is a more sensitive measure. The final result will be the

Proposition. *For a variety $X \subset \mathbb{P}_N$ and a smooth point $p \in X$, we denote by $|\mathbb{I}_p^*|$ the linear system of quadrics in $\mathbb{P}(T_p X)$ associated to the second fundamental form. Then for almost every p we have*

$$\delta_*(X) = \min \{\dim \operatorname{Sing} Q : Q \in |\mathbb{I}_p^*|\} + 1 \, .$$

In particular, $\delta_*(X) = 0$ iff there is a smooth point $p \in X$ and a smooth quadric Q in $|\mathbb{I}_p^*|$.

In addition, if we take a hyperplane H such that $\mathbb{T}_p X \subset H \subset \mathbb{P}_N$ with contact locus X_H and the corresponding quadric $Q \subset \mathbb{P}(T_p X)$ in $|\mathbb{I}_p^*|$ (2.4.4) then for almost every p and H we have

$$\dim \operatorname{Sing} Q + 1 = \dim X_H \, .$$

Of course, we have to define $\operatorname{Sing} \mathbb{P}(T_p X) = \mathbb{P}(T_p X)$ to include the extreme case $|\mathbb{I}_p^*| = \emptyset$ of a linear variety X.

The relation between dual defect and singular loci will be established via certain Hesse matrices which are derived from a local parameterization of X. This requires some preparations.

If $\widehat{X} \subset \mathbb{C}^{N+1}$ is the affine cone of dimension $n + 1$, and codimension $m = N - n$, we may assume

$$q = (1, 0, \dots, 0) \in \widehat{X}$$

is a smooth point over p. We choose a neighborhood U of

$$o = (0, \dots, 0) \in \mathbb{C}^{n+1}$$

with a local parameterization

$$\mathbb{C}^{n+1} \supset U \xrightarrow{\Phi} \widehat{X}, \quad t = (t_0, \dots, t_n) \longmapsto (1 + t_0, t_1, \dots, t_n, f_1(t), \dots, f_m(t)) \, ,$$

where $f_1, \dots, f_m$ are holomorphic functions such that

$$f_i(o) = \frac{\partial f_i}{\partial t_j}(o) = 0 \quad \text{for } i = 1, \dots, m, \quad j = 0, \dots, n \, .$$

This can be achieved by a linear change of coordinates in $\mathbb{C}^{N+1}$, bringing the tangent space of $\widehat{X}$ at q to the prescribed position $x_{n+1} = \dots = x_N = 0$.

Now we have the $(n + 1) \times (n + 1)$-Hesse matrices

$$A_i = \left(\frac{\partial^2 f_i}{\partial t_j \partial t_k}(o) \right)_{j,k} \quad \text{for } i = 1, \dots, m \,,$$

and with the identifications $T_q \widehat{X} = \mathbb{C}^{n+1}$, $N_q \widehat{X} = \mathbb{C}^m$ induced by the above choice of coordinates; according to formula $(*)$ in 2.4.1, the second fundamental form

$$\widehat{\mathbb{I}}_q : T_q \widehat{X} \times T_q \widehat{X} \longrightarrow N_q \widehat{X}$$

is described by $(A_1, \dots, A_m)$. Since $q \in K_q$ (2.4.3), we have

$$A_i = \begin{pmatrix} 0 \; 0 \cdots 0 \\ 0 \\ \vdots \quad B_i \\ 0 \end{pmatrix} \quad \text{for } i = 1, \dots, m \,,$$

and the matrices $B_1, \dots, B_m$ define the linear system $|\mathbb{I}_p^*|$ in $\mathbb{P}(T_p X) = \mathbb{P}_{n-1}$, where $p = (1 : 0 : \dots : 0) \in X$. According to 2.4.4, every normal direction

$$\lambda = (\lambda_1 : \dots : \lambda_m) \in \mathbb{N}_p(X) = \mathbb{P}_{m-1}$$

determines a quadric $Q_\lambda \subset \mathbb{P}_{n-1}$. In our computation, we obtain

$$A_\lambda := \sum_{i=1}^m \lambda_i A_i \,, \quad B_\lambda := \sum_{i=1}^m \lambda_i B_i \,, \quad Q_\lambda = \{ y \in \mathbb{P}_{n-1} : {}^t y B_\lambda y = 0 \} \,.$$

Since rank $B_\lambda = $ rank A_λ, we have

$$\dim \operatorname{Sing} Q_\lambda = n - \operatorname{rank} A_\lambda - 1 \,. \tag{$*$}$$

Recall that A_λ is the Hessian of $f_\lambda = \sum \lambda_i f_i$ at o.

2.5.4 Computation of the Dual Defect

We use the local parameterization of X given in 2.5.3 to describe the conormal bundle $C_X \subset X \times \mathbb{P}_N^*$ and its projection $\pi^* : C_X \longrightarrow \mathbb{P}_N^*$ [Kl, I-5]. More precisely, we consider the affine cone

$$\widehat{C}_X \subset \mathbb{C}^{N+1} \times (\mathbb{C}^{N+1})^* \quad \text{with coordinates } (x_0, \dots, x_N, a_0, \dots, a_N) \,.$$

If we examine the Jacobian of the map

$$\Phi : U \longrightarrow \widehat{X}$$

of 2.5.3 and the defining condition for C_X, we obtain a local parameterization

$$\Psi : U \times \mathbb{C}^m \longrightarrow \widehat{C}_X \,,$$

$$(t, \lambda) \longmapsto \left(\Phi(t), -\frac{\partial f_\lambda}{\partial t_0}(t), \ldots, -\frac{\partial f_\lambda}{\partial t_n}(t), \lambda_1, \ldots, \lambda_m \right) , \quad \text{where } f_\lambda := \sum_{i=1}^m \lambda_i f_i .$$

If $\widehat{\pi}^* : \widehat{C}_X \longrightarrow (\mathbb{C}^{N+1})^*$ denotes the projection, we consider

$$\xi := \widehat{\pi}^* \circ \Psi : U \times \mathbb{C}^m \longrightarrow (\mathbb{C}^{N+1})^* .$$

This map is given by the last $N + 1$ components of Ψ, and we can compute its Jacobian:

$$\text{Jac}_{(t,\lambda)} \xi = \left(\begin{array}{cc} -\text{Hess}_t f_\lambda & * \\ 0 & E_m \end{array} \right) .$$

This implies the useful formula

$$\text{rank Jac}_{\psi(t,\lambda)} \widehat{\pi}^* = \text{rank Hess}_t f_\lambda + m. \qquad (**)$$

For $t = o$, we have $p = (1 : 0 : \ldots : 0)$ and every hyperplane containing $\mathbb{T}_p X$ is of the form

$$H_\lambda = \{ x \in \mathbb{P}_N : \lambda_1 x_{n+1} + \ldots + \lambda_m x_N = 0 \} ,$$

with $\lambda \in \mathbb{C}^m$. Then $(**)$ specializes to

$$\text{rank Jac}_{(p,H_\lambda)} \pi^* = \text{rank Hess}_o f_\lambda + m - 1 .$$

The hyperplane H_λ corresponds to a direction $\lambda \in \mathbb{P}(N_p X) = \mathbb{P}_{m-1}$ and, hence, to a quadric Q_λ in the linear system $|\mathbb{I}_p^*|$; thus, formula $(*)$ of 2.5.3 implies

$$\text{rank Jac}_{(p,H_\lambda)} \pi^* = N - \dim \text{Sing } Q_\lambda - 2 . \qquad (***)$$

This formula can also be seen in terms of the contact locus: for $H \in X^*$, this is the linear space

$$X_H = \pi(\pi^{*-1}(H)) \subset X$$

(2.1.6). Since this is the fiber of π^*, we have

$$\text{rank Jac}_{(p,H)} \pi^* + \dim X_H = \dim C_X = N - 1$$

for almost all points $(p, H) \in C_X$, and $(***)$ implies

$$\dim X_{H_\lambda} = \dim \text{Sing } Q_\lambda + 1 ,$$

almost everywhere. Since

$$\delta_*(X) = \dim X_{H_\lambda}$$

for almost all p and λ, *Proposition 2.5.3 is proved*. In particular,

$$\dim X^* = N - 1$$

is equivalent to the existence of a single point $(p, H) \in C_X$ such that

$$\text{rank Jac}_{(p,H)} \pi^* = N - 1 .$$

The above computation also yields a description of the contact locus in local coordinates:
if
$$p = (1 : 0 : \ldots : 0) \quad \text{and} \quad H = (a_0 : \ldots : a_N) \supset \mathbb{T}_p X,$$
then $a_0 = \ldots = a_n = 0$; hence, $H = H_\lambda$ for some $\lambda = (\lambda_1 : \ldots : \lambda_m)$ and

$$\Phi^{-1}(\widehat{X}_{H_\lambda}) = V \left(f_\lambda, \frac{\partial f_\lambda}{\partial t_0}, \ldots, \frac{\partial f_\lambda}{\partial t_n} \right) \subset U \subset \mathbb{C}^{n+1}.$$

2.5.5 The Surface Case

Recall that a variety $X \subset \mathbb{P}_N$ is

- ordinary if $\delta_*(X) = 0$, i.e. $\dim X^* = N - 1$, and

- nondevelopable if $\delta_\gamma(X) = 0$, i.e. $\dim \gamma(X) = \dim X$.

By 2.3.3, we have $\delta_\gamma(X) \leq \delta_*(X)$; hence, every ordinary variety is non developable. The Segre varieties $X_{m,n}$ with $m < n$ are an example that the converse need not be true (2.5.1). The simplest example is the threefold $X_{1,2}$. However, for small dimension the previous characterizations of δ_* and δ_γ yield the

Proposition. *A nondevelopable variety $X \subset \mathbb{P}_N$ with* $\dim X \leq 2$ *is ordinary.*

Proof. For $\dim X = 1$, the only nonordinary or developable varieties are lines (2.1.7).

For $\dim X = 2$, we use Propositions 2.5.2 and 2.5.3. Thus, our claim is a direct consequence of the following BERTINI type result for linear systems of quadrics in $\mathbb{P}(T_p X) \cong \mathbb{P}_1$.
$\square$

Lemma. *Assume we have a linear system $\mathbb{E}$ of quadrics on $\mathbb{P}_1$ which are all singular. Then* $\dim \mathbb{E} = 0$. *In particular, the quadrics in $\mathbb{E}$ are all singular in the same point.*

Proof of the Lemma. The complete linear system of quadrics in $\mathbb{P}_1$ is isomorphic to $\mathbb{P}_2$ with coordinates $(a : b : c)$, where the corresponding equation is

$$ax_0^2 + 2bx_0 x_1 + cx_1^2.$$

The quadric is singular, i.e. a double point, iff $b^2 - ac = 0$. Thus, the singular quadrics in $\mathbb{P}_1$ lie on a smooth quadric $C \subset \mathbb{P}_2$. The only possible linear subspaces in C are points. $\square$

Since every developable surface which is not a plane has singularities, we obtain the following corollary.

Corollary. *Every smooth and nonplane surface $X \subset \mathbb{P}_N$ is ordinary (i.e. $\dim X^* = N - 1$).*

2.5.6 Classification of Developable Hypersurfaces

As we have seen in 2.3.3, for a hypersurface $X \subset \mathbb{P}_N$ the Gauß image $\gamma(X)$ and the dual variety $X^\vee$ in $\mathbb{G}(N - 1, N)$ are the same. Furthermore, $X = \gamma(X)^\vee$, i.e. the hypersurface can easily be reconstructed from its Gauß image. Conversely, if we start with some variety $Y \subset \mathbb{P}_N^\vee$, then $X := Y^\vee \subset \mathbb{P}_N$ is a hypersurface, iff Y was ordinary, and in this case $Y = \gamma(X)$. In particular, the dimension of Y is equal to the Gauß rank. This implies the

Classification Theorem for Developable Hypersurfaces. *Hypersurfaces* $X \subset \mathbb{P}_N$ *of Gauß rank* r *are classified by ordinary varieties* $Y \subset \mathbb{P}_N^\vee$ *of dimension* r *in the following sense:*

$$X = Y^\vee \quad and \quad Y = \gamma(X) \,.$$

Unfortunately, the classification of ordinary (or equivalently of nonordinary) varieties seems to be difficult in general [E]. But for $r \leq 2$ we have, as a consequence of 2.2.8 and 2.5.5:

Proposition. *Take a variety* $Y \subset \mathbb{P}_N$, $\dim Y \leq 2$. *If* $\dim Y = 1$, *then*

$$Y \text{ is nonordinary} \Longleftrightarrow Y \text{ is a line} \,.$$

If $\dim Y = 2$, *then*

$$Y \text{ is nonordinary} \Longleftrightarrow Y \text{ is either a plane, a cone, or a tangent scroll} \,.$$

For these nonordinary surfaces the dual varieties are easy to describe (2.5.7).

Corollary. *Hypersurfaces* $X \subset \mathbb{P}_N$ *of Gauß rank* 2 *are in one to one correspondence with surfaces* $Y \subset \mathbb{P}_N^\vee$ *except planes, cones, or tangent scrolls via duality:*

$$X = Y^\vee \quad and \quad Y = X^\vee = \gamma(X) \,.$$

It seems to be an interesting question, how the structure of the surface Y is reflected in the hypersurface X. This would make the "classification" much more explicit.

The classification of varieties of arbitrary dimension but Gauß rank 1 is established in 2.2.8 and 2.3.3. The general problem of classification for arbitrary dimension and Gauß rank seems to be difficult. At the moment there are two approaches: The first is to show that a developable variety is the union of smaller developable varieties. Griffiths and Harris proved that a developable variety with one-dimensional Gauß fibers is the union of two-dimensional tangent scrolls and cones [GH₂]. This was refined by Akivis and Goldberg to the statement that an arbitary developable variety is the union of osculating scrolls, cones, or developable hypersurfaces [AG₂]. The second approach is to try to understand the relation between the developable variety and its focal variety. This lead to the classification of developable hypersurfaces of Gauß rank 2 in $\mathbb{P}_4$ by Akivis, Goldberg and Landsberg [AGL]. This result will be improved to a classification of arbitary developable varieties of Gauß rank 2 in [P₃].

2.5.7 Dual Defect of Uniruled Varieties

One may suspect that a ruling of a sufficiently high dimension causes a positive dimensional contact locus, hence a dual defect. More precisely, we have a bound which is attained by the Segre varieties (2.5.1):

Proposition. *If a variety $X \subset \mathbb{P}_N$ is uniruled by k-planes, then*

$$\delta_*(X) \geq 2k - \dim X .$$

Proof. For every $p \in X$, we have a linear space L of dimension k such that $p \in L \subset X$. For $0 \neq q \in \widehat{X}_{\mathrm{sm}}$ and every quadric $Q \in \widehat{\mathbb{II}}_q^*$, we know, from 2.4.5, that

$$Q(\widehat{L}) = 0 .$$

We may choose a basis (e_i) of $\mathbb{C}^{N+1}$ such that

$$\widehat{L} = \mathrm{span}\,\{e_0, \dots, e_k\} \subset T_q\widehat{X} = \mathrm{span}\,\{e_0, \dots, e_n\}$$

with $n = \dim X$. Then every $Q \in \widehat{\mathbb{II}}_q^*$ is given by an $(n+1)$-square matrix of the form

$$A_Q = \left(\begin{array}{c|c} 0 & * \\ \hline * & * \end{array} \right),$$

where the 0 is a $(k+1)$-square matrix. This implies

$$\mathrm{rank}\, A_q \leq 2(n-k)$$

for every $Q \in \widehat{\mathbb{II}}_q^*$; hence, we have

$$\dim \mathrm{Sing}\, Q \geq n+1 - 2(n-k) = 2k - n + 1 .$$

For the corresponding quadric $\overline{Q} \subset \mathbb{P}(T_pX)$ this implies

$$\dim \mathrm{Sing}\, \overline{Q} \geq 2k - n - 1 .$$

So the assertion follows from 2.5.3.   □

The above proof also shows why the ruling does not necessarily produce a Gauß defect: the form of the matrices A_Q gives only a common bound for the dimension of the singular loci of the quadrics Q, but their position may be variable (compare this to 2.5.2).

2.5.8 Varieties with Very Small Dual Varieties

If the dual variety $X^\vee$ is a point then X is a hyperplane. There are some other cases of varieties X which can easily be reconstructed from their very small dual $X^\vee$. By 2.2.6 it is sufficient to consider the case where $X^\vee$ is nondegenerate. In the simplest case, $X^\vee = C$ is a curve. As we have seen in 2.1.7 (Corollary 2),

$$X = \mathrm{Tan}^{\,(N-2)}\widetilde{C}\,,$$

i.e. X is the osculating scroll of order $N-2$ of the dual curve $\widetilde{C} \subset \mathbb{P}_N$. In particular, X is a hypersurface.

Next, we consider the case where the dual variety $X^\vee$ is a surface. If $X^\vee$ is ordinary then X is a hypersurface of Gauß rank 2 (2.5.6). If $X^\vee$ is nonordinary, the interesting case is that of a tangent scroll. To simplify the notation we dualize the situation.

Proposition. *If $X = \mathrm{Tan}\, C$ for some nondegenerate curve $C \subset \mathbb{P}_N$ and if $\widetilde{C} \subset \mathbb{P}_N^\vee$ denotes the dual curve (2.1.7), then*

$$X^\vee = \mathrm{Tan}^{\,(N-3)}\widetilde{C}\,.$$

In particular, codim $X^\vee = 2$.

Proof. We use the notations and results of 2.1.7. If

$$\varphi : S \longrightarrow C \subset \mathbb{P}_N$$

is a parameterization with a compact Riemann surface S, we have the curves

$$\varphi^{(k)} : S \longrightarrow C^{(k)} \subset \mathbb{G}(k, N) \quad \text{for } 0 \le k \le N-1 \text{ and}$$

$$\widetilde{\varphi} = \mathcal{D} \circ \varphi^{(N-1)} : S \longrightarrow \widetilde{C} \subset \mathbb{P}_N^\vee\,.$$

Since the tangent plane to a tangent scroll is the osculating plane of the curve, we have

$$\mathbb{T}_p X = \varphi^{(2)}(s) \quad \text{for almost every } p \in \varphi^{(1)}(s) \text{ and } s \in S\,.$$

Furthermore,

$$\mathcal{D} \circ \varphi^{(2)}(s) = \{H \in \mathbb{P}_N^\vee : H \supset \mathbb{T}_p X,\, p \in \varrho^{(1)}(s)\}\,; \quad \text{hence,}$$

$$Y^\vee = \bigcup_{s \in S}(\mathcal{D} \circ \varphi^{(2)})(s) = \bigcup_{s \in S}\widetilde{\varphi}^{(N-3)}(s) = \mathrm{Tan}^{(N-3)}\widetilde{C}\,. \qquad \square$$

As an exercise the reader may show that the dual of a nondegenerate two-dimensional cone is the intersection of some $\mathrm{Tan}^{(N-2)}C$ with a hyperplane.

Chapter 3

Tangent and Secant Varieties

Natural examples for varieties swept out by linear spaces are the tangent and secant variety of an arbitrary variety, i.e. the unions of tangents and secants. The contributions of F.L. Zak have brought dramatic progress in this very classical part of projective algebraic geometry. Our last chapter is devoted to this topic and will lead to open questions.

3.1 Zak's Theorems

ZAK's Theorems are one of the main tools to study tangent and secant varieties. They give estimates on the dimension of the tangent and secant varieties. These global statements will complement our local techniques of Chapter 2. We start with the necessary definitions.

3.1.1 Tangent Spaces, Tangent Cones, and Tangent Stars

At a smooth point of a variety $X \subset \mathbb{P}_N$ there is no question of what the tangent space should be. We simply have to decide if we want to use the affine version $T_p X$ or the embedded one $\mathbb{T}_p X$ (1.1.1). At a singular point however, there exist several possibilities for embedded tangent objects. We will use three of them:

1. The *projective (embedded Zariski-) tangent space* $\mathbb{T}_p X$

$$\mathbb{T}_p X := \{\mathbb{C} v \in \mathbb{P}_N : \operatorname{grad}_p F(v) = 0 \quad \text{for all } F \in I(X)\}.$$

   Hence, $\mathbb{T}_p X$ is given by the linear terms of the polynomials in $I(X)$ developed around p. Instead of considering the whole ideal $I(X)$, it is enough to take a generating set of $I(X)$.

2. The *tangent cone* $\mathbb{T}_p^c X$

   Analogous to the tangent space $\mathbb{T}_p X$, we define $\mathbb{T}_p^c X$ as the variety in $\mathbb{P}_N$ which is defined by the initial terms of the polynomial of $I(X)$ developed around p. Here it is not enough to consider only a generating set of $I(X)$. Obviously, $\mathbb{T}_p^c X$ is contained in $\mathbb{T}_p X$.

   There are two more geometric descriptions of the tangent cone:

   (a) The tangent cone is the union of all limit lines of sequences $\overline{p x_i}$ where $x_i \in X$ is a sequence of points converging to p.

(b) The tangent cone is the union of all embedded tangent lines to arcs in X through p at p. Here we use the fact that for an arc there is a well-defined tangent line (1.5.2).

Since we will not use the tangent cone often, we refer the reader to [H, Lecture 20] for the proof of the above statements and more information. In analytic geometry, different kinds of tangents were considered in [Wh₂].

3. The *tangent star* $\mathbb{T}_p^\star X$

The tangent star is the union of all limit lines of sequences $\overline{x_i y_i}$ where x_i, $y_i \in X$ are sequences of points converging to p.

These three tangent objects are related by inclusions.

Proposition. *For $p \in X \subset \mathbb{P}_N$,*

$$\mathbb{T}_p^c X \subset \mathbb{T}_p^\star X \subset \mathbb{T}_p X .$$

Proof. The first inclusion is clear. To see the second, let x_i, $y_i \to p$ be sequences that converge to p, such that the sequence of lines $\overline{x_i y_i}$ has a limit l. We have to show that this limit line l lies in $\mathbb{T}_p X$. We compute on the affine cone $\widehat{X}$ and choose $q \in \widehat{p} \setminus \{0\}$, $\widetilde{x}_i \in \widehat{x}_i \setminus \{0\}$, $\widetilde{y}_i \in \widehat{y}_i \setminus \{0\}$, such that $\widetilde{x}_i, \widetilde{y}_i \to q$. Further, we pick any norm $\| \ \|$ on $\mathbb{C}^{N+1}$ and choose $v_i \in \mathbb{C}^N$ and $\lambda_i \in \mathbb{C}$ such that

$$\lambda_i v_i := \widetilde{x}_i - \widetilde{y}_i \quad \text{and} \quad \|v_i\| = 1 .$$

Clearly, $\lambda_i \to 0$. After passing to a subsequence, we may assume that the v_i converge to some v, then the affine cone of the limit line $\widehat{l}$ is spanned by v and q. For any homogeneous $f \in I(X)$ the following equations hold

$$
\begin{aligned}
0 = f(\widetilde{x}_i) &= f(\widetilde{y}_i + \lambda_i v_i) \\
&= f(\widetilde{y}_i) + \lambda_i \operatorname{grad}_{\widetilde{y}_i} f(v_i) + \lambda_i^2 \cdots \\
&= 0 + \lambda_i \operatorname{grad}_{\widetilde{y}_i} f(v_i) + \lambda_i^2 \cdots .
\end{aligned}
$$

Dividing by λ_i gives

$$\operatorname{grad}_{\widetilde{y}_i} f(v_i) + \lambda_i \cdots = 0 .$$

Since $\lambda_i \to 0$, $v_i \to v$, $\widetilde{y}_i \to q$, we conclude

$$\operatorname{grad}_q f(v) = 0 .$$

Together with $\operatorname{grad}_q f(q) = \deg f \cdot f(q) = 0$, by EULER's formula, this implies that the limit line $l = \mathbb{P}(\operatorname{span}\{q, v\})$ lies in $\mathbb{T}_p X$. $\qquad \square$

If the point $p \in X$ is smooth, then all tangent objects are the same. The following example shows that in the singular case they may all be different.

Example. To keep the example simple, we will use a reducible algebraic set X, for which the above tangent objects can be defined in the same way. Let $X \subset \mathbb{P}_3$ be the union of three lines that intersect in a point p, but that are not contained in any plane. The tangent cone $\mathbb{T}_p^c X$ is X itself. The tangent star $\mathbb{T}_p^\star X$ is the union of the three planes that are spanned by any two lines. Finally, the tangent space is the whole $\mathbb{P}_3$.

3.1.2 Zak's Theorem on Tangent and Secant Varieties

So far we have considered different spaces for every point of a variety. Now we take their union over all points.

In 2.2.7, we had defined the tangent variety of a curve. If $X \subset \mathbb{P}_N$ is an arbitrary variety of dimension n, then we have a Gauß image $\gamma(X) \subset \mathbb{G}(n, N)$, and

$$\mathrm{Tan}\, X := \bigcup_{T \in \gamma(X)} T \subset \mathbb{P}_N$$

is a variety (2.1.1), which is uniruled by $\gamma(X)$ if it has the expected dimension $\dim \gamma(X) + n$. The variety $\mathrm{Tan}\, X$ is called the *tangent variety* of X.

The union of the tangent stars,

$$\mathrm{Tan}^\star X := \bigcup_{p \in X} \mathbb{T}_p^\star X \subset \mathbb{P}_N,$$

is called the *tangent star variety* of X. Finally,

$$\mathrm{Sec}\, X := X \# X \subset \mathbb{P}_N,$$

the join of X with itself (2.1.3), is called the *secant variety* of X.

Proposition. *For a variety $X \subset \mathbb{P}_N$, the spaces defined above are in the following relation:*

$$\mathrm{Tan}\, X = \bigcup_{p \in X} \mathbb{T}_p^c X \subset \mathrm{Tan}^\star X \subset \mathrm{Sec}\, X \subset \mathbb{P}_N.$$

Proof. The first equality is clear. On the one hand, $\mathrm{Tan}\, X$ is the closure of

$$\bigcup_{p \in X_{\mathrm{sm}}} \mathbb{T}_p X = \bigcup_{p \in X_{\mathrm{sm}}} \mathbb{T}_p^c X,$$

so it is contained in $\bigcup_{p \in X} \mathbb{T}_p^c X$.

On the other hand, by the remark (a) in 3.1.1, the tangent cones are the limits of lines that lie in $\bigcup_{p \in X_{\mathrm{sm}}} \mathbb{T}_p X$, so the reverse inclusion also holds.

The inclusion $\mathrm{Tan}\,X \subset \mathrm{Tan}^{\star}X$ follows from the Proposition 3.1.1.

For the remaining assertions, we use the *join correspondence*

$$S_X := \text{closure of } \{(x, y, p) \in (X \times X \setminus \Delta_X) \times \mathbb{P}_N : p \in \overline{xy}\} \subset X \times X \times \mathbb{P}_N\,,$$

where $\Delta_X \subset X \times X$ is the diagonal, with the canonical projection $\pi : S_X \to \mathbb{P}_N$. As in 2.1.3, it can be seen that S_X is a variety, and obviously,

$$\mathrm{Sec}\,X = \pi(S_X) \subset \mathbb{P}_N\,.$$

By 3.1.1, we have

$$\mathrm{Tan}^{\star}X = \pi(S_X \cap \Delta_X \times \mathbb{P}_N) \subset \pi(S_X)\,.$$

Hence, the tangent star variety is algebraic, but not necessarily irreducible.   □

For a smooth point $p \in X$, we know $\mathbb{T}_p X = \mathbb{T}_p^{\star}X$; nevertheless, $\mathrm{Tan}^{\star}X$ may be much bigger than $\mathrm{Tan}\,X$.

Example. Let $C \subset \mathbb{P}_4$ be a smooth nondegenerate curve and $p \in \mathbb{P}_4 \setminus C$. We define $X :=$ $C \# p$. The secant variety $\mathrm{Sec}\,X$ is $p \# \mathrm{Sec}\,C$. This is easy to see: any point of $\mathrm{Sec}\,X$ lies on a line intersecting two lines $\overline{pc}$ and $\overline{pc'}$ with $c, c' \in C$, i.e. $\mathrm{Sec}\,X$ is the union of planes $\overline{pcc'}$. Hence, $\mathrm{Sec}\,X$ is the union of cones over secant lines of C, so $\mathrm{Sec}\,X = p \# \mathrm{Sec}\,C$.

Because of $\dim \mathrm{Sec}\,C = 3$, we find that $\dim \mathrm{Sec}\,X = \dim p \# \mathrm{Sec}\,C = 4$. Therefore, $\mathrm{Sec}\,X$ is the whole space $\mathbb{P}_4$.

On the other hand, to compute $\mathrm{Tan}\,X$ recall from 2.2.5 that the tangent planes to X are constant along the punctured line $\overline{pc} \setminus \mathrm{Sing}\,X$. Thus, the dimension of $\mathrm{Tan}\,X$ is only 3.

Every line through the vertex p of the cone X is the limit of secants approaching p, consequently

$$\mathbb{T}_p^{\star}X = \mathbb{P}_4\,, \quad \text{hence } \mathrm{Tan}^{\star}X = \mathbb{P}_4\,, \quad \text{so } \mathrm{Tan}\,X \neq \mathrm{Tan}^{\star}X\,.$$

For technical reasons, we need relative versions of the above constructions.

For $Y \subset X \subset \mathbb{P}_N$, we define the following sets

$$S_{Y,X} \qquad = \text{closure of } \{(y, x, p) \in (Y \times X \setminus \Delta) \times \mathbb{P}_N : p \in \overline{xy}\} \subset Y \times X \times \mathbb{P}_N$$

$$\mathrm{Sec}\,(Y, X) \ = \pi(S_{Y,X}) = Y \# X$$

$$\mathbb{T}_y^{\star}(Y, X) \ = \pi(\{y\} \times \{y\} \times \mathbb{P}_N \cap S_{Y,X})$$

$$\mathrm{Tan}^{\star}(Y, X) = \pi(\Delta_Y \times \mathbb{P}_N \cap S_{Y,X}) = \bigcup_{y \in Y} \mathbb{T}_y^{\star}X, \quad \text{where } \Delta_Y = \{(y, y) \in Y \times Y\} \cong Y\,.$$

Hence, the relative tangent star $\mathbb{T}_y^{\star}(Y, X)$ is the union of the limit lines of sequences $\overline{y_n x_n}$ where $y_n \in Y$, $x_n \in X$ are sequences that converge to $y \in Y$.

The essential theorem about the secant and tangent star varieties is

ZAK's **Theorem.** *For varieties $Y \subset X \subset \mathbb{P}_N$ one of the following statements holds*

1. $\dim \operatorname{Tan}^\star(Y, X) = \dim Y + \dim X$ *and* $\dim \operatorname{Sec}(Y, X) = \dim Y + \dim X + 1$

2. $\operatorname{Tan}^\star(Y, X) = \operatorname{Sec}(Y, X)$.

The proof uses FULTON and HANSEN's

Connectedness Theorem. *Let $f : X \to \mathbb{P}_m \times \mathbb{P}_m$ be a finite morphism from a projective variety of dimension n. If $n > m$, then the inverse image $f^{-1}(\Delta)$ of the diagonal $\Delta \subset \mathbb{P}_m \times \mathbb{P}_m$ is connected.*

Idea of proof. We use the rational map

$$l : \qquad \mathbb{P}_m \times \mathbb{P}_m \qquad \dashrightarrow \qquad \mathbb{P}_{2m+1}$$

$$(s_0 : \ldots : s_m; t_0 : \ldots : t_m) \longmapsto (s_0 : \ldots : s_m : t_0 : \ldots : t_m)$$

to linearize $\mathbb{P}_m \times \mathbb{P}_m$. Note that l is biregular on Δ and $L = l(\Delta) \subset \mathbb{P}_{2m+1}$ is a linear subspace of dimension m. Applying a BERTINI type theorem to $l \circ f : X \to \mathbb{P}_{2m+1}$ we conclude that $(l \circ f)^{-1}(L) = f^{-1}(\Delta)$ is connected. For details see [FL].   □

Proof of ZAK's *theorem.* Since $\operatorname{Tan}^\star(Y, X) \subset \operatorname{Sec}(Y, X)$ and $\operatorname{Sec}(Y, X)$ is irreducible, we only need to check the dimensions.

Let $m := \dim \operatorname{Tan}^\star(Y, X)$ and assume the theorem is false, i.e. $m < \dim X + \dim Y$ and $m < \dim \operatorname{Sec}(Y, X)$. We choose a linear subspace $L \subset \mathbb{P}_N$ of dimension $N - m - 1$ that does not intersect $\operatorname{Tan}^\star(Y, X)$. On the other hand, L must intersect $\operatorname{Sec}(Y, X)$ in at least a point, i.e. there exist $y \in Y, x \in X, x \neq y$ with $\overline{xy} \cap L \neq \emptyset$. Denoting the projection of $\mathbb{P}_N$ from L by π, we obtain a finite morphism

$$f := \pi|_Y \times \pi|_X : Y \times X \longrightarrow \mathbb{P}_m \times \mathbb{P}_m .$$

By our choice of y and x, $f(y, x) = f(y, y)$. Since the inverse image of Δ is connected by the Connectedness Theorem, there is an arc (y_t, x_t) in $f^{-1}(\Delta)$ with $(y_0, x_0) = (y, y)$ and $(y_1, x_1) = (y, x)$. (Actually, there is only a chain of arcs with this property, but the changes needed in the argument are obvious.) For a pair (y_t, x_t), $y_t \neq x_t$, being in $f^{-1}(\Delta)$ means that the line $\overline{y_t x_t}$ intersects L. Thus, we have an arc of lines which intersects L for $y_t \neq x_t$. Hence, its limit line for $t = 0$, which is a line of the tangent star $\mathbb{T}_y^\star(Y, X)$, also intersects L, contradicting the choice of L.   □

Corollary. *For a variety $X \subset \mathbb{P}_N$, one of the following statements holds*

1. $\dim \operatorname{Tan}^\star X = 2 \dim X$ *and* $\dim \operatorname{Sec} X = 2 \dim X + 1$

2. $\operatorname{Tan}^\star X = \operatorname{Sec} X$.

The corollary is false if we replace $\operatorname{Tan}^\star X$ by $\operatorname{Tan} X$.

3.1.3 Theorem on Tangencies

Very often ZAK's Theorem is used in the following form:

Theorem. *Let $Z \subset Y \subset X \subset \mathbb{P}_N$ be a chain of varieties and $L \subset \mathbb{P}_N$ be a linear subspace with $\dim L \geq \dim X$, but $L \not\supset X$ such that*

$$\mathbb{T}_y^\star(Y, X) \subset L \quad for \ y \in Y \setminus Z \, .$$

Then

$$\dim Y \leq \dim L - \dim X + \dim Z + 1 \, .$$

Proof. We reduce to the case $Z = \emptyset$ by intersecting Z, Y, X, L with a general codimension $\dim Z + 1$ linear subspace M. We only need to note that

1. It doesn't matter for the inequality.

2. $Y \cap M$ and $X \cap M$ are irreducible by BERTINI's Theorem.

3. If $X \not\subset L$ then $X \cap M \not\subset L \cap M$.

4. If $\mathbb{T}_y^\star(Y, X) \subset L$ for $y \in Y \cap M$ then

$$\mathbb{T}_y^\star(Y \cap M, X \cap M) \subset \mathbb{T}_y^\star(Y, X) \cap M \subset L \cap M \, .$$

Now we are in the situation that on the one hand,

$$\mathrm{Tan}^\star(Y, X) \subset L \, ,$$

so $\dim \mathrm{Tan}^\star(Y, X) \leq \dim L$. On the other hand, $X \subset \mathrm{Sec}\,(Y, X) \not\subset L$. Therefore, $\mathrm{Tan}^\star(Y, X) \neq \mathrm{Sec}\,(Y, X)$ and ZAK's Theorem yields

$$\dim \mathrm{Tan}^\star(Y, X) = \dim Y + \dim X \, .$$

Combining this with $\dim \mathrm{Tan}^\star(Y, X) \leq \dim L$ proves the theorem. □

Using $\mathbb{T}_y^\star(Y, X) \subset \mathbb{T}_y^\star X \subset \mathbb{T}_y X$ and setting $Z = \mathrm{Sing}\, X$, we can weaken the theorem to

Corollary (ZAK's Theorem on Tangencies). *Let $X \subset \mathbb{P}_N$ be a variety and $L \subset \mathbb{P}_N$ be a linear subspace with $\dim L \geq \dim X$, but $L \not\supset X$, then*

$$\dim\{p \in X_{\mathrm{sm}} : \mathbb{T}_p X \subset L\} \leq \dim L - \dim X + \dim \mathrm{Sing}\, X + 1 \, .$$

If we take a tangent space $\mathbb{T}_p X$, $p \in X_{\mathrm{sm}}$, for L and recall the Linearity Theorem (2.3.2), we obtain the

Corollary. *For a variety $X \subset \mathbb{P}_N$, the general fiber of the Gauß map is a linear subspace of dimension at most $\dim \mathrm{Sing}\, X + 1$.*

In particular, for a smooth variety, the Gauß map is birational.

3.2   Third and Higher Fundamental Forms

So far we have used only the second fundamental form, and it might be surprising how many invariants are determined by it. The secant variety Sec X is the first example where third order invariants of X must be considered to obtain even the simplest type of information like its dimension.

3.2.1 Definition

Recall from 2.4.1, that we defined the second fundamental form first on an affine variety $Y \subset \mathbb{C}^{N+1}$, dim $Y = n+1$, and then moved this definition to the projective case $X \subset \mathbb{P}_N$ by considering the affine cone $\widehat{X} \subset \mathbb{C}^{N+1}$. We will do the same here for the third fundamental form.

For a smooth point $q \in Y \subset \mathbb{C}^{N+1}$, we define the *second osculating space* by

$$T_q^{(2)} Y := \operatorname{span} \{\operatorname{Im} \mathbb{I}_q, T_q Y\} .$$

There will be an open subset U of Y where these osculating spaces have constant dimension, say $a + n + 1$. U is the domain of definition for the second Gauß map

$$\gamma^{(2)} : Y \dashrightarrow \operatorname{Gr}(n + a + 1, N + 1), \quad q \longmapsto T_q^{(2)} Y .$$

Differentiating this at a point of definition $q \in U$ and using the canonical isomorphism of 1.4.2, yields

$$d_q \gamma^{(2)} : T_q Y \longrightarrow T_{T_q^{(2)} Y} \operatorname{Gr}(n + a + 1, N + 1) = \operatorname{Hom}(T_q^{(2)} Y, \mathbb{C}^{N+1} / T_q^{(2)} Y) .$$

Lemma. *For any $v \in T_q Y$, the linear map*

$$d_q \gamma^{(2)}(v) : T_q^{(2)} Y \longrightarrow \mathbb{C}^{N+1} / T_q^{(2)} Y$$

has $T_q Y$ in its kernel.

Proof. We must use the description of the canonical isomorphism of 1.4.2. Let $\alpha : S \to Y$ be a holomorphic path with $\alpha(o) = q$ and $\alpha'(o) = v$ and $\varrho : S \to \mathbb{C}^{N+1}$ be any moving vector with $\varrho \in \gamma \circ \alpha \subset \gamma^{(2)} \circ \alpha$. We set $w = \varrho(o) \in T_q Y$. Then on the one hand,

$$d_q \gamma^{(2)}(v)(w) = \varrho'(o) + T_q^{(2)} Y ,$$

and on the other hand, by 2.4.1

$$\mathbb{I}_q(v, w) = d_q \gamma(v)(w) = \varrho'(o) + T_q Y \subset T_q^{(2)} Y ,$$

therefore

$$d_q \gamma^{(2)}(v)(w) = \varrho'(o) + T_q^{(2)} Y = 0 + T_q^{(2)} Y . \qquad \square$$

Because of the lemma $d_q\gamma^{(2)}(v)$ descends to a linear map

$$d_q\gamma^{(2)}(v) : T_q^{(2)}Y/T_qY \longrightarrow \mathbb{C}^{N+1}/T_q^{(2)}Y,$$

and we can compose it with the second fundamental form to obtain the *third fundamental form*

$$\begin{aligned}
\mathbb{III}_q : \quad T_qY \times T_qY \times T_qY \quad &\longrightarrow \quad \mathbb{C}^{N+1}/T_q^{(2)}Y \\
(v, w, u) \quad &\longmapsto \quad d_q\gamma^{(2)}(v)(\mathbb{II}_q(w, u)).
\end{aligned}$$

This expression for $\mathbb{III}_q$ appears rather complicated, but in local coordinates $\mathbb{III}_q$ is what one expects (see proof of the next Proposition).

The map $\mathbb{III}_q$ has properties analogous to $\mathbb{II}_q$.

Proposition. $\mathbb{III}_q$ *is symmetric.*

Proof. We want to derive an expression for $\mathbb{III}_q(v, w, u)$ similar to the expression for $\mathbb{II}_q$ in the proof of Proposition 2.4.1. There it was shown that if $\Phi : U \to Y \subset \mathbb{C}^{N+1}$ is a parameterization of Y with $\Phi(o) = q$ then

$$\mathbb{II}_q(w, u) = \sum_{i,j} w_i \frac{\partial^2\Phi}{\partial t_i \partial t_j}(o)u_j + T_qY,$$

where $w = \sum w_i \frac{\partial\Phi}{\partial t_i}(o)$ and $u = \sum u_j \frac{\partial\Phi}{\partial t_j}(o)$.

We choose a path $\beta : S \to U$, $\beta(o) = o$, such that for $\alpha := \Phi \circ \beta$ we have $\alpha(o) = q$ and $\alpha'(o) = v = \sum v_k \frac{\partial\Phi}{\partial t_k}(o)$. Further, we extend $w, u \in T_qY$ to vector fields $w(s), u(s)$ along $\alpha(s)$. Then

$$s \longmapsto \mathbb{II}_{\alpha(s)}(w(s), u(s)) \in T_{\alpha(s)}^{(2)}Y/T_{\alpha(s)}Y$$

can be used to compute

$$\begin{aligned}
\mathbb{III}_q(v, w, u) &= \frac{\partial}{\partial s}\left(\mathbb{II}_{\alpha(s)}(w(s), u(s))\right)(o) + T_q^{(2)}Y \\
&= \frac{\partial}{\partial s}\left(\sum_{i,j} w_i(s)\frac{\partial^2\Phi}{\partial t_i \partial t_j}(\beta(s))u_j(s)\right)(o) + T_q^{(2)}Y \\
&= \sum_{i,j} w_i'(o)\frac{\partial^2\Phi}{\partial t_i \partial t_j}(o)u_j + \sum_{i,j,k} w_i\frac{\partial^3\Phi}{\partial t_i \partial t_j \partial t_k}(o)v_k u_j \\
&\quad + \sum_{i,j} w_i\frac{\partial^2\Phi}{\partial t_i \partial t_j}(o)u_j'(o) + T_q^{(2)}Y \\
&= \sum_{i,j,k} v_k w_i u_j\frac{\partial^3\Phi}{\partial t_i \partial t_j \partial t_k}(o) + T_q^{(2)}Y
\end{aligned}$$

This expression is clearly symmetric in v, w, u. $\square$

The value of $\mathrm{I\!I\!I}_q(v, v, v)$ is particularly easy to compute.

Lemma. *If $Y \subset \mathbb{C}^{N+1}$ and $\alpha : S \to Y$, $\alpha(o) = q$, is a holomorphic path, then*

$$\mathrm{I\!I\!I}_q(\alpha'(o), \alpha'(o), \alpha'(o)) = \alpha'''(o) + T_q^{(2)}Y\,.$$

Proof. We may take

$$s \longmapsto \mathrm{I\!I}_{\alpha(s)}\left(\alpha'(s), \alpha'(s)\right) = \alpha''(s) + T_{\alpha(s)}Y \in T_{\alpha(s)}^{(2)}Y / T_{\alpha(s)}Y$$

as a moving point for the computation of $\mathrm{I\!I\!I}_q(\alpha'(o), \alpha'(o), \alpha'(o))$ and obtain

$$\mathrm{I\!I\!I}_q(\alpha'(o), \alpha'(o), \alpha'(o)) = d_q\gamma^{(2)}\left(\alpha'(o)\right)\left(\mathrm{I\!I}_q(\alpha'(o), \alpha'(o))\right)$$

$$= \frac{\partial}{\partial s}\left(\mathrm{I\!I}_{\alpha(s)}(\alpha'(s), \alpha'(s))\right)(o) + T_q^{(2)}Y$$

$$= \frac{\partial}{\partial s}(\alpha''(s))(o) + T_q^{(2)}Y = \alpha'''(0) + T_q^{(2)}Y\,. \qquad \square$$

By polarization, this cubic map determines the trilinear map $\mathrm{I\!I\!I}_q$.

One should note that this formula is independent of the second derivative of α. This makes $\mathrm{I\!I\!I}_q$ easy to compute, but it also means a loss of information, so $\mathrm{I\!I\!I}_q$ is complemented by the introduction of F_3. We refer the reader to [L_1] for the definition and applications of F_3.

It remains to treat the special case when Y is a cone.

Lemma. *If $Y \subset \mathbb{C}^{N+1}$ is a cone and $q \in Y$, then*

$$\mathrm{I\!I\!I}_q(q, T_qY, T_qY) = \mathrm{I\!I\!I}_q(T_qY, q, T_qY) = \mathrm{I\!I\!I}_q(T_qY, T_qY, q) = 0\,.$$

Proof. This follows immediately from the definition and $\mathrm{I\!I}_q(q, T_qY) = 0$. $\qquad \square$

We transfer this construction to the projective case $X \subset \mathbb{P}_N$ by applying it to the affine cone $\widehat{X} \subset \mathbb{C}^{N+1}$ and obtain the third fundamental form for $q \in \widehat{X}$

$$\widehat{\mathrm{I\!I\!I}}_q : \mathrm{Sym}^3 T_q\widehat{X} \longrightarrow \mathbb{C}^{N+1} / T_q^{(2)}\widehat{X}\,.$$

The definition of the higher fundamental forms is now obvious, so we will only fix the notation here. The *k-th fundamental form* is

$$\mathrm{F\!F}^{(k)} : \mathrm{Sym}^k T_qY \longrightarrow \mathbb{C}^{N+1} / T_q^{(k-1)}Y\,,$$

where $T_q^{(k-1)}Y$ is the $(k-1)$-th osculating space, i.e.

$$T_q^{(k-1)}Y := \mathrm{span}\,\{\mathrm{Im}\,\mathrm{F\!F}^{(k-1)}, T_q^{(k-2)}Y\}\,.$$

In particular, $\mathrm{I\!I} = \mathrm{F\!F}^{(2)}$ and $\mathrm{I\!I\!I} = \mathrm{F\!F}^{(3)}$.

3.2.2 Vanishing of Fundamental Forms

If the second fundamental form of a variety $X \subset \mathbb{P}_N$ vanishes, then X is a linear space. Equivalently, we may say that X lies in the tangent space at a general point of X. In this form, the statement generalizes to the higher fundamental forms: if the k-th fundamental form vanishes, X lies in the $(k-1)$-th osculating space at a general point of X.

To prove this, we start with a lemma.

Lemma. *If the k-th fundamental form of a variety vanishes, then all higher ones as well.*

Proof. By induction it is enough to show $\mathbb{FF}^{(k+1)} = 0$. Let $\Phi : U \subset \mathbb{C}^{n+1} \to \widehat{X}$ be a local parameterization of $\widehat{X}$ near $\Phi(o) = q$. Then

$$0 = \mathbb{FF}^{(k)}_{\Phi(t)}\left(\frac{\partial \Phi}{\partial t_{i_1}}(t), \ldots, \frac{\partial \Phi}{\partial t_{i_k}}(t)\right) = \frac{\partial^k \Phi}{\partial t_{i_1} \ldots \partial t_{i_k}}(t) + T^{(k-1)}_{\Phi(t)}\widehat{X}.$$

Since $T^{(k-1)}_{\Phi(t)}\widehat{X}$ is spanned by the derivatives of Φ of order less than or equal to $k-1$, the expression $\frac{\partial^k \Phi}{\partial t_{i_1} \ldots \partial t_{i_k}}(t)$ can be written as a linear combination of those over the holomorphic function ring. Differentiating with respect to $t_{i_{k+1}}$ yields that $\frac{\partial^{k+1}\Phi}{\partial t_{i_1} \ldots \partial t_{i_{k+1}}}(t)$ is a linear combination of the derivatives of Φ of order less than or equal to k, i.e. at $t = o$ we find that

$$\frac{\partial^{k+1}\Phi}{\partial t_{i_1} \ldots \partial t_{i_{k+1}}}(o) \in T^{(k)}_q\widehat{X}.$$

Therefore,

$$\mathbb{FF}^{(k+1)}_q\left(\frac{\partial \Phi}{\partial t_{i_1}}(o), \ldots, \frac{\partial \Phi}{\partial t_{i_{k+1}}}(o)\right) = \frac{\partial^{k+1}\Phi}{\partial t_{i_1} \ldots \partial t_{i_k}}(o) + T^{(k)}_q\widehat{X} = 0 + T^{(k)}_q\widehat{X}.$$

Because $\mathbb{FF}^{(k+1)}$ is zero on the basis vectors, it vanishes.   □

Proposition. *Let $X \subset \mathbb{P}_N$ be a variety and $k \geq 2$. Then*

$$\mathbb{FF}^{(k)} = 0 \iff X \subset \mathbb{P}(T^{(k-1)}_q\widehat{X}) \quad \text{for a general } q \in \widehat{X}.$$

In particular, if the second fundamental form of a variety vanishes, then the variety is a linear space.

Proof. The "$\Leftarrow$" direction of the theorem is trivial; we must only note that the condition on the right hand side is true for almost all $q \in \widehat{X}$, since it is true for a general point.

To prove the reverse direction, we choose an arbitrary path α through q, $\alpha(o) = q$. Then for all $l \geq k$, by the assumption and the above lemma,

$$0 = \mathbb{FF}^{(l)}_q(\alpha'(o), \ldots, \alpha'(o)) = \alpha^{(l)}(o) + T^{(l-1)}_q\widehat{X}.$$

From $\mathbb{FF}_q^{(l)} = 0$ for $l \geq k$, it follows by the definition of $T_q^{(l)}\widehat{X}$ that

$$T_q^{(k-1)}\widehat{X} = T_q^{(k)}\widehat{X} = T_q^{(k+1)}\widehat{X} = \ldots .$$

Therefore,

$$\alpha^{(l)}(o) \in T_q^{(k-1)}\widehat{X} \quad \text{for all } l \in \mathbb{N},$$

so that the expansion of $\alpha(s)$ is

$$\alpha(s) = q + (s-o)\alpha'(o) + \tfrac{1}{2}(s-o)^2\alpha''(o) + \tfrac{1}{3!}(s-o)^3\alpha^{(3)}(o) + \ldots \quad \text{with } \alpha^{(l)}(o) \in T_q^{(k-1)}X.$$

We see that the path α lies inside $T_q^{(k-1)}\widehat{X}$. Because α was an arbitrary path through $q \in \widehat{X}$, the whole $\widehat{X}$ must lie inside $T_q^{(k-1)}\widehat{X}$. $\square$

3.3 Tangent Varieties

3.3.1 The Dimension of the Tangent Variety

We start our examination of the tangent variety,

$$\operatorname{Tan} X = \text{closure of} \bigcup_{p \in X_{\mathrm{sm}}} \mathbb{T}_p X,$$

by computing its dimension. Since the tangent variety for an n-dimensional variety $X \subset \mathbb{P}_N$ is the union of an n-dimensional family of n-dimensional linear subspaces, we expect it to have the dimension $2n$, so we define the *tangent defect* as

$$\delta_{\mathrm{T}} := 2n - \dim \operatorname{Tan} X .$$

It is determined by the second fundamental form of X.

Theorem. *For a variety $X \subset \mathbb{P}_N$, let $p \in X$ be a general point, $q \in \widehat{p} \setminus \{0\}$, and $v \in T_q\widehat{X}$ a general tangent vector. Then*

$$\dim \operatorname{Tan} X = \dim X + \dim \widehat{\mathbb{I}}_q(v, T_q\widehat{X}) .$$

Thus, the tangent defect is

$$\delta_{\mathrm{T}} = \dim X - \dim \widehat{\mathbb{I}}_q(v, T_q\widehat{X}) = \dim \operatorname{Ker} \widehat{\mathbb{I}}_q(v, \_) - 1 .$$

Proof. Let $q \in \widehat{X}$ be a general point and $\Phi : U \subset \mathbb{C}^{n+1} \to \widehat{X}$ a local parameterization with $\Phi(o) = q$. Then $(\operatorname{Tan} X)^\wedge$ is locally the image of

$$\Psi : \quad U \times \mathbb{C}^{n+1} \quad \longrightarrow \quad \mathbb{C}^{N+1}$$

$$(t, \lambda) \quad \longmapsto \quad \sum_{i=0}^{n} \lambda_i \frac{\partial \Phi}{\partial t_i}(t) .$$

At a general smooth point $v := \Psi(o, \mu)$, the tangent space to $(\mathrm{Tan}\, X)^\wedge$ is given by the span of the columns of $d\Psi$, thus

$$T_v(\mathrm{Tan}\, X)^\wedge = \mathrm{span}\left\{ \frac{\partial \Phi}{\partial t_j}(o), \sum_{i=0}^{n} \mu_i \frac{\partial^2 \Phi}{\partial t_i \partial t_j}(o) : j = 0, \dots, n \right\}.$$

According to formula $(*)$ in 2.4.1,

$$\widehat{\mathrm{II}}_q \left(v, \frac{\partial \Phi}{\partial t_j}(o) \right) = \sum_{i=0}^{n} \mu_i \frac{\partial^2 \Phi}{\partial t_i \partial t_j}(o) + T_q \widehat{X};$$

hence,

$$T_v(\mathrm{Tan}\, X)^\wedge = \widehat{\mathrm{II}}_q(v, T_q \widehat{X}) + T_q \widehat{X}.$$

In particular, $\dim(\mathrm{Tan}\, X)^\wedge = n + 1 + \dim \widehat{\mathrm{II}}_q(v, T_q \widehat{X})$. $\square$

By the last formula in the above proof, we see that

$$T_v(\mathrm{Tan}\, X)^\wedge = \widehat{\mathrm{II}}_q(v, T_q \widehat{X}) + T_q \widehat{X} = \widehat{\mathrm{II}}_q(\alpha v, T_q \widehat{X}) + T_q \widehat{X} = T_{\alpha v}(\mathrm{Tan}\, X)^\wedge$$

for any $\alpha \in \mathbb{C}^*$. This implies the following proposition.

Proposition. *Let $X \subset \mathbb{P}_N$ be a variety, $p \in X$ a general smooth point, and L a line in $\mathbb{T}_p X$ through p. Then the tangent spaces to $\mathrm{Tan}\, X$ are constant along $L \cap (\mathrm{Tan}\, X)_{\mathrm{sm}}$.*

3.3.2 Developability of the Tangent Variety

As we have just seen, tangent varieties are always developable. If X is a curve, then the uniruling of developability of $\mathrm{Tan}\, X$ is given by the tangent spaces to X. This is wrong if the dimension of X is greater than 1. Then the uniruling of developability of $\mathrm{Tan}\, X$ is only a subruling of the uniruling of $\mathrm{Tan}\, X$ given by the tangent spaces. In general it is a one-dimensional uniruling, but if the tangent variety is degenerate, it is developable of degree ≥ 2. GRIFFITHS and HARRIS considered this the deepest result of their paper [GH2, 5.17]. We will prove LANDSBERG's improved version [L2, 12.1] in an elementary way.

Theorem. *For a projective variety $X \subset \mathbb{P}_N$, the Gauß defect of its tangent variety is at least its tangent defect plus one,*

$$\delta_\gamma(\mathrm{Tan}\, X) \geq \delta_\mathrm{T} + 1.$$

In particular, every tangent variety is developable.

Proof. In the beginning, we follow the proof of the last section, only with stronger assumptions on the choice of coordinates. Let $\Phi : U \subset \mathbb{C}^{n+1} \to \widehat{X}$ be a local parameterization of $\widehat{X}$ with $\Phi(o) = q$, where $o = (1, 0, \dots, 0)$. We may assume that the parameterization respects the cone structure in such a way that changes in the t_0-coordinate induce movements along the lines of the cone, i.e. $\Phi(t_0, t_1, \ldots, t_n) = t_0 \Phi(1, t_1, \ldots, t_n)$. We assume further that the coordinates are chosen in such a way that $v := \frac{\partial \Phi}{\partial t_1}(o) \in T_q \widehat{X}$ is a general vector, so

$$\dim \operatorname{Ker} \widehat{\mathbb{I}}_q (v, \_) = \delta_T + 1 .$$

We already know that

$$\widehat{\mathbb{I}}_q \left(v, \frac{\partial \Phi}{\partial t_0}(o) \right) = \widehat{\mathbb{I}}_q (v, \Phi(o)) = \widehat{\mathbb{I}}_q(v, q) = 0 ,$$

by a final adjustment of the coordinates we can achieve that

$$\operatorname{Ker} \widehat{\mathbb{I}}_q (v, \_) = \operatorname{span} \left\{ \frac{\partial \Phi}{\partial t_0}(o), \frac{\partial \Phi}{\partial t_{n-\delta_T+1}}(o), \ldots, \frac{\partial \Phi}{\partial t_n}(o) \right\} .$$

Now we again consider the map

$$\widetilde{\Psi} : \quad U \times \mathbb{C}^{n+1} \quad \longrightarrow \quad \mathbb{C}^{N+1}$$

$$(t, \lambda) \quad \longmapsto \quad \sum_{i=0}^{n} \lambda_i \frac{\partial \Phi}{\partial t_i}(t) .$$

Its image is a part of $(\operatorname{Tan} X)^\wedge$ and $\widetilde{\Psi}(o, e_1) = v$. At the point (o, e_1) its differential is

$$\left(\frac{\partial^2 \Phi}{\partial t_0 \partial t_1}(o), \ldots, \frac{\partial^2 \Phi}{\partial t_n \partial t_1}(o), \frac{\partial \Phi}{\partial t_0}(o), \ldots, \frac{\partial \Phi}{\partial t_n}(o) \right) .$$

By the assumptions on the coordinates, the vectors

$$\frac{\partial \Phi}{\partial t_0}(o), \ldots, \frac{\partial \Phi}{\partial t_n}(o), \frac{\partial^2 \Phi}{\partial t_1 \partial t_1}(o), \ldots, \frac{\partial^2 \Phi}{\partial t_{n-\delta_T} \partial t_1}(o)$$

are linearly independent and span the column space of the matrix above which is the tangent space

$$T_v (\operatorname{Tan} X)^\wedge = \widehat{\mathbb{I}}_q (v, T_q \widehat{X}) + T_q \widehat{X} .$$

We conclude that

$$\Psi(t_1, \ldots, t_{n-\delta_T}, \lambda) := \widetilde{\Psi}(1, t_1, \ldots, t_{n-\delta_T}, 0, \ldots, 0, \lambda)$$

parameterizes $(\operatorname{Tan} X)^\wedge$ locally.

To prove the theorem, it will be enough to show that at the point (o, e_1)

$$\operatorname{span} \left\{ \frac{\partial \Psi}{\partial \lambda_0}, \frac{\partial \Psi}{\partial \lambda_1}, \frac{\partial \Psi}{\partial \lambda_{n-\delta_T+1}}, \ldots, \frac{\partial \Psi}{\partial \lambda_n} \right\} \subset \operatorname{Sing} \widehat{\mathbb{I}}^*_{\operatorname{Tan} X, v} .$$

Here and further on, we suppress evaluation at (o, e_1) in the formulas for the ease of notation.

We compute for the index ranges

$$i, j = 0 \dots, n$$

$$l = 1, \dots, n - \delta_{\text{Tan}}$$

$$k = 0, n - \delta_{\text{Tan}} + 1, \dots, n .$$

$$\widehat{I\!I}_{\text{Tan}\,X,v} \left(\frac{\partial \Psi}{\partial \lambda_i}, \frac{\partial \Psi}{\partial \lambda_j} \right) = \frac{\partial^2 \Psi}{\partial \lambda_i \partial \lambda_j} + T_v(\text{Tan}\,X)^{\wedge} = 0$$

$$\widehat{I\!I}_{\text{Tan}\,X,v} \left(\frac{\partial \Psi}{\partial \lambda_1}, \frac{\partial \Psi}{\partial t_l} \right) = \frac{\partial^2 \Psi}{\partial \lambda_1 \partial t_l} + T_v(\text{Tan}\,X)^{\wedge}$$

$$= \frac{\partial^2 \Phi}{\partial t_1 \partial t_l}(o) + T_v(\text{Tan}\,X)^{\wedge} = 0$$

$$\widehat{I\!I}_{\text{Tan}\,X,v} \left(\frac{\partial \Psi}{\partial \lambda_k}, \frac{\partial \Psi}{\partial t_l} \right) = \frac{\partial^2 \Psi}{\partial \lambda_k \partial t_l} + T_v(\text{Tan}\,X)^{\wedge}$$

$$= \frac{\partial^2 \Phi}{\partial t_k \partial t_l}(o) + T_v(\text{Tan}\,X)^{\wedge}$$

$$= \widehat{I\!I}_q \left(\frac{\partial \Phi}{\partial t_k}(o), \frac{\partial \Phi}{\partial t_l}(o) \right) + \left(\widehat{I\!I}_q(v, T_q \widehat{X}) + T_q \widehat{X} \right) = 0 ,$$

where the last term vanishes by the following lemma.   □

Lemma. *Let* $\widehat{I\!I} : T \times T \to N$ *be a symmetric bilinear form and* $v \in T$ *general, i.e.* $W_v := \text{Ker}\,(\widehat{I\!I}(v, \_) : T \to N)$ *has minimal dimension among all* W_u, $u \in T$. *Then*

$$\widehat{I\!I}(W_v, T) \subset \widehat{I\!I}(v, T) .$$

Proof. Let $u \in T$ be any vector and $w \in W_v$. We have to show

$$\widehat{I\!I}(w, u) \in \widehat{I\!I}(v, T) .$$

Pick any arc $\alpha : U \subset \mathbb{C} \to T$ with $\alpha(o) = v$ and $\alpha'(o) = u$. If $w : U \to T$, $w(o) = w$, is a moving point for the moving spaces $t \mapsto W_{\alpha(t)}$, which we may assume to be all of the same minimal dimension, then

$$\widehat{I\!I}(w(t), \alpha(t)) = 0 .$$

We differentiate this at the point $t = o$ to get

$$\widehat{I\!I}(w(o), \alpha'(o)) + \widehat{I\!I}(w'(o), \alpha(o)) = 0$$

$$\Longrightarrow \widehat{I\!I}(w, u) = -\widehat{I\!I}(w'(o), v) = -\widehat{I\!I}(v, w'(o)) \in \widehat{I\!I}(v, T) .$$ □

3.3.3 Singularities of the Tangent Variety

It was already known by CAYLEY that the tangent scroll of a curve is singular along the curve. Naturally, the analogous statement holds for higher dimensions.

Proposition. *Let $X \subset \mathbb{P}_N$ be such that the tangent variety* Tan X *is not a linear space, then* $X \subset$ Sing Tan X.

Proof. Because of the closeness of Sing Tan X, we need only to prove that an open subset of X is contained in Sing Tan X.

We take a general point p of X and consider the tangent spaces to Tan X along $\mathbb{T}_p X \cap$ (Tan X)$_{\mathrm{sm}}$. We shall divide the proof in two cases:

1. these tangent spaces are all the same (special case, but it exists, e.g. X a curve)

2. they are not (general case).

The second case can be treated like the cone case (2.2.5) because the tangent spaces to Tan X are all the same along a line in $\mathbb{T}_p X$ through p (3.3.1).

In the first case, we reduce the problem to the case where Tan X is a hypersurface. Obviously, the Proposition is true iff it is true for all projections of Tan X that project Tan X to a hypersurface, i.e. projections from a linear space $L \subset \mathbb{P}_N$ where L is of dimension codim Tan $X - 2$ and $L \cap$ Tan $X = \emptyset$.

For the following computation, we again use a parameterization $\Phi : U \to \widehat{X}$ with $\Phi(t_0, \dots, t_n) = t_0 \Phi(1, t_1, \dots, t_n)$ and the map

$$\Psi : \quad U \times \mathbb{C}^{n+1} \quad \longrightarrow \quad \mathbb{C}^{N+1}$$

$$(t, \lambda) \quad \longmapsto \quad \sum_{i=0}^{n} \lambda_i \frac{\partial \Phi}{\partial t_i}(t),$$

whose image is a part of (Tan X)$^\wedge$. By assumption, the tangent spaces,

$$T_{\Psi(t,\lambda)}(\mathrm{Tan}\, X)^\wedge = \mathrm{span} \left\{ \frac{\partial \Phi}{\partial t_j}(t), \sum_{i=0}^{n} \lambda_i \frac{\partial^2 \Phi}{\partial t_i \partial t_j}(t) : j = 0, \dots, n \right\},$$

are all the same for smooth points $\Psi(t, \lambda)$ with fixed t, but arbitrary λ. Thus we have

$$T_{\Psi(t,\lambda)}(\mathrm{Tan}\, X)^\wedge = \mathrm{span} \left\{ \frac{\partial \Phi}{\partial t_i}(t), \frac{\partial^2 \Phi}{\partial t_i \partial t_j}(t) : i, j = 0, \dots, n \right\} = T_{\Phi(t)}^{(2)} \widehat{X}. \qquad (*)$$

Let F be an equation for the hypersurface $(\operatorname{Tan} X)^\wedge \subset \mathbb{C}^{N+1}$. The linear subspace $T_{\Phi(t)}\widehat{X}$ lies in this hypersurface, so at the point $\Phi(t) = \Psi(t, t_0 e_0) \in (\operatorname{Tan} X)^\wedge$

$$\operatorname{grad}_{\Phi(t)} F(v) = 0$$

$$\operatorname{Hess}_{\Phi(t)} F(v, w) = 0$$

$$\operatorname{tri}_{\Phi(t)} F(v, w, u) = 0 \quad \text{for all } v, w, u \in T_{\Phi(t)}\widehat{X},$$

where tri denotes the third derivative. Suppressing evaluation at t in the notation, we differentiate $\operatorname{Hess}_\Phi F(\frac{\partial \Phi}{\partial t_i}, \frac{\partial \Phi}{\partial t_j}) = 0$ with respect to t_k and obtain

$$0 = \operatorname{tri}_\Phi F(\tfrac{\partial \Phi}{\partial t_i}, \tfrac{\partial \Phi}{\partial t_j}, \tfrac{\partial \Phi}{\partial t_k}) + \operatorname{Hess}_\Phi F(\tfrac{\partial^2 \Phi}{\partial t_i \partial t_k}, \tfrac{\partial \Phi}{\partial t_j}) + \operatorname{Hess}_\Phi F(\tfrac{\partial \Phi}{\partial t_i}, \tfrac{\partial^2 \Phi}{\partial t_j \partial t_k})$$

$$= \operatorname{Hess}_\Phi F(\tfrac{\partial^2 \Phi}{\partial t_i \partial t_k}, \tfrac{\partial \Phi}{\partial t_j}) + \operatorname{Hess}_\Phi F(\tfrac{\partial \Phi}{\partial t_i}, \tfrac{\partial^2 \Phi}{\partial t_j \partial t_k}).$$

If we also take into account the equations with the permuted indices, we get

$$\operatorname{Hess}_\Phi F \left(\frac{\partial^2 \Phi}{\partial t_i \partial t_j}, \frac{\partial \Phi}{\partial t_k} \right) = 0.$$

From $(*)$ we know

$$0 = \operatorname{grad}_\Phi F \left(\frac{\partial^2 \Phi}{\partial t_i \partial t_j} \right);$$

differentiating with respect to t_k, yields

$$0 = \operatorname{Hess}_\Phi F(\tfrac{\partial^2 \Phi}{\partial t_i \partial t_j}, \tfrac{\partial \Phi}{\partial t_k}) + \operatorname{grad}_\Phi F(\tfrac{\partial^3 \Phi}{\partial t_i \partial t_j \partial t_k})$$

$$= \operatorname{grad}_\Phi F(\tfrac{\partial^3 \Phi}{\partial t_i \partial t_j \partial t_k}).$$

Hence, together with $(*)$, we get

$$\operatorname{grad}_{\Phi(t)} F(v) = 0 \quad \text{for } v \in \operatorname{span} \left\{ \frac{\partial \Phi}{\partial t_i}(t), \frac{\partial^2 \Phi}{\partial t_i \partial t_j}(t), \frac{\partial^3 \Phi}{\partial t_i \partial t_j \partial t_k}(t) \right\} = T^{(3)}_{\Phi(t)}\widehat{X}.$$

Because $\dim T^{(2)}_{\Phi(t)}\widehat{X} = \dim T_{\Psi(t,\lambda)}(\operatorname{Tan} X)^\wedge = N$, the derivative $\operatorname{grad}_{\Phi(t)} F$ is zero and $(\operatorname{Tan} X)^\wedge$ is singular in $\Phi(t)$ if $T^{(2)}_{\Phi(t)}\widehat{X} \subsetneq T^{(3)}_{\Phi(t)}\widehat{X}$.

Let us assume that $\operatorname{Tan} X$ is smooth at a general point of X, then $T^{(2)}_{\Phi(t)}\widehat{X} = T^{(3)}_{\Phi(t)}\widehat{X}$, i.e. $\mathrm{III} = 0$. By 3.2.2, $\widehat{X}$ lies in the linear subspace $T^{(2)}_{\Phi(t)}\widehat{X}$. Hence,

$$(\operatorname{Tan} X)^\wedge \subset T^{(2)}_{\Phi(t)}\widehat{X} = T_{\Phi(t)}(\operatorname{Tan} X)^\wedge.$$

Thus, $(\operatorname{Tan} X)^\wedge$ is a linear subspace, which contradicts the assumptions.   □

Corollary. *The osculating scroll* $\operatorname{Tan}^{(k)} C$ *of a curve* C *is either a linear space or singular along* $\operatorname{Tan}^{(k-1)} C$.

Proof. This follows from the Proposition by $\operatorname{Tan}^{(k)} C = \operatorname{Tan}(\operatorname{Tan}^{(k-1)} C)$.   □

3.4 The Dimension of the Secant Variety

If X is a smooth variety, then its tangent star variety equals its tangent variety. In view of ZAK's theorem, the dimension of the secant variety is either $2 \dim X + 1$ or equals the dimension of the tangent variety, which was easy to compute. We would like to have a criterion for the degeneracy of the secant variety. We will show a necessary condition here. Equivalent conditions for the degeneracy of the secant variety are more difficult to give; in fact, the ones that we have found in the literature up to now are false.

Theorem. *Let $X \subset \mathbb{P}_N$ be a smooth variety with a degenerate secant variety. Then the third fundamental form vanishes, $\widehat{\mathrm{I\!I\!I}} = 0$.*

In particular, X lies in $\mathbb{T}_p^{(2)} X$ for any general $p \in X$.

Note that if the tangent variety is degenerate, which is easy to check, then the secant variety is also degenerate, hence $\widehat{\mathrm{I\!I\!I}} = 0$.

Proof. The "in particular" follows from 3.2.2. To prove the main statement, we use again a parameterization $\Phi : U \subset \mathbb{C}^{n+1} \to \widehat{X}$ with $\Phi(o) = q$ for $o = (1, 0, \dots, 0)$ and $\Phi(t_0, t_1, \dots, t_n) = t_0 \Phi(1, t_1, \dots, t_n)$. By TERRACINI's Lemma, the tangent space at any smooth point z on the line $\overline{\Phi(o)\Phi(o + \alpha v)}$ for $\alpha \in \mathbb{C}^*$ and $v \in \mathbb{C}^{n+1}$ general, is given by $T_{\Phi(o)}\widehat{X} + T_{\Phi(o+\alpha v)}\widehat{X}$, so

$$T_z(\mathrm{Sec}\, X)^\wedge = \mathrm{span}\left\{ \frac{\partial \Phi}{\partial t_0}(o), \dots, \frac{\partial \Phi}{\partial t_n}(o), \frac{\partial \Phi}{\partial t_0}(o + \alpha v), \dots, \frac{\partial \Phi}{\partial t_n}(o + \alpha v) \right\}.$$

We assume that the coordinates are chosen such that $v = e_1$. We expand $\frac{\partial \Phi}{\partial t_i}(o + \alpha e_1)$ into a Taylor series, where we suppress the evaluation at o for better readability.

$$\frac{\partial \Phi}{\partial t_0}(o + \alpha e_1) = \Phi(o + \alpha e_1) = \Phi + \alpha \frac{\partial \Phi}{\partial t_1} + \frac{1}{2}\alpha^2 \frac{\partial^2 \Phi}{(\partial t_1)^2} + \frac{1}{3!}\alpha^3 \frac{\partial^3 \Phi}{(\partial t_1)^3} + \dots$$

$$\frac{\partial \Phi}{\partial t_1}(o + \alpha e_1) = \frac{\partial \Phi}{\partial t_1} + \alpha \frac{\partial^2 \Phi}{(\partial t_1)^2} + \frac{1}{2}\alpha^2 \frac{\partial^3 \Phi}{(\partial t_1)^3} + \dots$$

$$\frac{\partial \Phi}{\partial t_j}(o + \alpha e_1) = \frac{\partial \Phi}{\partial t_j} + \alpha \frac{\partial^2 \Phi}{\partial t_j \partial t_1} + \frac{1}{2}\alpha^2 \frac{\partial^3 \Phi}{\partial t_j (\partial t_1)^2} + \dots \quad \text{for } j = 2, \dots, n.$$

Hence, because $\alpha \neq 0$, the tangent space $T_z(\mathrm{Sec}\, X)^\wedge$ is also the span of

$$\frac{\partial \Phi}{\partial t_0}, \dots, \frac{\partial \Phi}{\partial t_n}, \left(\frac{1}{3!} - \frac{1}{2}\frac{1}{2}\right)\frac{\partial^3 \Phi}{(\partial t_1)^3} + \left(\frac{1}{4!} - \frac{1}{2}\frac{1}{3!}\right)\alpha \frac{\partial^4 \Phi}{(\partial t_1)^4} + \dots,$$

$$\frac{\partial^2 \Phi}{(\partial t_1)^2} + \frac{1}{2}\alpha \frac{\partial^3 \Phi}{(\partial t_1)^3} + \dots, \quad \frac{\partial^2 \Phi}{\partial t_j \partial t_1} + \frac{1}{2}\alpha \frac{\partial^3 \Phi}{\partial t_j (\partial t_1)^2} + \dots \quad \text{for } j = 2, \dots, n.$$

Letting α tend to zero, we conclude that for a z on the limit line $\lim_{\alpha \to 0} \overline{\Phi(o)\Phi(o + \alpha v)}$, which is a tangent line of $\widehat{X}$, the tangent space $T_z(\mathrm{Sec}\, X)^\wedge$ contains the following vectors:

$$\frac{\partial \Phi}{\partial t_0}, \dots, \frac{\partial \Phi}{\partial t_n}, \frac{\partial^2 \Phi}{\partial t_1 \partial t_1}, \dots, \frac{\partial^2 \Phi}{\partial t_1 \partial t_n}, \frac{\partial^3 \Phi}{(\partial t_1)^3}.$$

The first $2n + 1$ vectors span $T_z(\operatorname{Tan} X)^\wedge = \widehat{\mathbb{II}}_q(\frac{\partial \Phi}{\partial t_1}, T_q \widehat{X}) + T_q \widehat{X}$; hence,

$$T_z(\operatorname{Tan} X)^\wedge + \mathbb{C}\frac{\partial^3 \Phi}{(\partial t_1)^3} \subset T_z(\operatorname{Sec} X)^\wedge. \qquad (*)$$

Now by ZAK's Theorem, the degeneracy of the secant variety implies $\operatorname{Tan} X = \operatorname{Sec} X$, so

$$T_z(\operatorname{Tan} X)^\wedge = T_z(\operatorname{Sec} X)^\wedge.$$

From $(*)$, we conclude

$$\frac{\partial^3 \Phi}{(\partial t_1)^3} \in T_z(\operatorname{Tan} X)^\wedge = \widehat{\mathbb{II}}_q\left(\frac{\partial \Phi}{\partial t_1}, T_q \widehat{X}\right) + T_q \widehat{X} \subset T_q^{(2)} \widehat{X}.$$

Therefore,

$$\widehat{\mathbb{III}}_q\left(\frac{\partial \Phi}{\partial t_1}, \frac{\partial \Phi}{\partial t_1}, \frac{\partial \Phi}{\partial t_1}\right) = \frac{\partial^3 \Phi}{(\partial t_1)^3} + T_q^{(2)} \widehat{X} = 0,$$

and, since $\frac{\partial \Phi}{\partial t_1}$ is a general vector, $\widehat{\mathbb{III}}_q = 0$. □

Bibliography

[AG₁] Akivis, M. and V. Goldberg. *Local Equivalence of Examples of Sacksteder and Bourgain.* Preprint, arXiv: math.DG/0002092.

[AG₂] Akivis, M. and V. Goldberg. *On the Structure of Submanifolds with Degenerate Gauss Mappings.* Preprint, arXiv: math.DG/0002093.

[AGL] Akivis, M., V. Goldberg and J. Landsberg. *On the structure of varieties with degenerate Gauss mappings.* Preprint, arXiv: math.AG/9908079.

[C] Cayley, A. *On Certain Developable Surfaces.* Quarterly Journal of Pure and Applied Mathematics, Volume VI (1864), pp. 108–126.

[D] Dombrowski, P. *Krümmungsgrößen gleichungsdefinierter Untermannigfaltigkeiten Riemannscher Mannigfaltigkeiten.* Math. Nachr. **38** (1968), pp. 133–180.

[E] Ein, L. *Varieties with Small Dual Varieties*, I. Invent. math. **86** (1986), pp.63–74.

[Ei] Eisenhart, L. *An Introduction to Differential Geometry with Use of the Tensor Calculus.* Princeton Univ. Press, 1940.

[F₁] Fischer, G. *Complex Analytic Geometry.* LNM 538. Springer, Berlin 1976.

[F₂] Fischer, G. *Plane Algebraic Curves.* American Mathematical Society 2001.

[FH] Fulton, W. and J. Hansen. *A connectedness theorem for projective varieties, with applications of intersections and singularities of mappings.* Ann. Math. (2) **110**, No. 1 (1979), pp. 159–166.

[FL] Fulton, W. and R. Lazarsfeld. *Connectivity and its applications in algebraic geometry.* Algebraic Geometry (Chicago, 1980), LNM 862, Springer, New York 1981, pp. 26–92.

[FOV] Flenner, H., L. O'Carroll and W. Vogel. *Joins and Intersections.* Springer, New York 1999.

[FW] Fischer, G. and H. Wu. *Developable Complex Analytic Submanifolds.* Int. J. Math. **6**, No. 2 (1995), pp. 229–272.

[G] Gauß, C. F. *Disquisitiones generales circa superficies curvas.* Collected works, Vol. 4, 1828.

[GH₁] Griffiths, P. and J. Harris. *Principles of Algebraic Geometry.* Wiley-Interscience, New York 1978.

[GH₂] Griffiths, P. and J. Harris. *Algebraic Geometry and Local Differential Geometry.* Annales scientifiques de L' École Normale Superérieure, 4e série, tome 12, No. 3 (1979), pp. 355–432.

[H]   Harris, J. *Algebraic Geometry.* Springer, New York 1993.

[HN] Hartman, P. and L. Nirenberg. *On spherical image maps whose Jacobians do not change signs.* Amer. Jour. Math. **81** (1959), pp. 901–920.

[Kl]   Kleiman, S. L. *Tangency and Duality.* Canadian Mathematical Society, Conference Proceedings, Volume 6 (1986). AMS Providence, Rhode Island 1986, pp. 163–225.

[Kli] Klingenberg, W. *A Course in Differential Geometry.* Springer, New York 1978.

[L₁]   Landsberg, J. *Algebraic Geometry and Projective Differential Geometry.* Research Institute of Mathematics - Global Analysis Research Center, Lecture Notes Series No. 45. Seoul National University 1999.

[L₂]   Landsberg, J. *On degenerate secant and tangential varieties and local differential geometry.* Duke **85**, No. 3 (1996), pp. 605–634.

[M]   Massey, W. *Surfaces of Gaussian Curvature Zero in Euclidean 3-space.* Tôhoku Math. J. (2) **14** (1962), pp. 73–79.

[Mu] Mumford, D. *Algebraic Geometry I: Complex projective varieties.* Springer, Berlin 1976.

[P₁]   Piontkowski, J. *A normal form for curves in Grassmannians.* Manuscripta Mathematica **89** (1996), pp. 79–85.

[P₂]   Piontkowski, J. *Abwickelbare Varietäten.* Dissertation, Düsseldorf 1995.

[P₃]   Piontkowski, J. *Developable Varieties of Gauss rank 2.* Preprint, 2001.

[Po]   Pogorelov, A. *Extension of the Theorem of Gauß of the Spherical Image of Surfaces of Bounded Extrinsic Curvature.* Dokl. Akad. Nauk **111** (1956), pp. 945–947.

[R]   Ran, Z. *The structure of Gauß-like maps.* Composito **52** (1984), pp. 171–172.

[S]   Sacksteder, R. *On hypersurfaces with no negative sectional curvature.* Amer. J. Math. **82**, No. 3 (1960), pp. 609–630.

[Sh]   Shafarevich, I. R. *Basic Algebraic Geometry.* Second Edition in 2 Volumes. Springer, Berlin 1994.

[V]   Vitter, A. *Twisted-Cylinder Theorem for Complex Submanifolds.* Preprint, 1979.

[W]   Wu, H. H. *Complete Developable Submanifolds in Real and Complex Euclidean Space.* Int. J. Math. **6**, No. 3 (1995), pp. 461–489.

[Wi]   Wilczynski, E. J. *Projective differential geometry of curves and ruled surfaces.* Chelsea Publ. Co., New York 1961.

[Wh$_1$]   Whitney, H. *Complex Analytic Varieties.* Addison-Wesley, Reading 1972.

[Wh$_2$]   Whitney, H. *Local Properties of Analytic Varieties* in S. Cairns *Differential and Combinatorial Topology.* Princeton Univ. Press, 1965.

[WZ]   Wu, H. H. and F. Zheng. *On Complete Developable Submanifolds in Complex Euclidean Space.* Preprint, 2000.

[Z]   Zak, F. *Tangents and Secants of Algebraic Varieties.* AMS, Providence, Rhode Island 1993.

Index

adapted parameterization, 64
affine cone, 13
algebraic set, 12
algebraic variety, 12
annihilator, 28, 33
associated curves, 45, 57, 76

base locus, see second fundamental form
Bertini's Theorem, 57, 115

Cayley method, 59
Classification Theorem, 77, 116
cone, 2, 10, 12, 69–72
conormal bundle, 54
contact locus, 56
Cramer's Rule, 19
critical directrix, 7, 10, 68
curvature, 99
cylinder, 2
Cylinder Theorem, 8

degeneracy map, 97, 103
degeneracy space, see second fundamental form
degenerate variety, 72
derived curves, 40
Developability Criterion, 65, 82
developable hypersurface, 116
developable ruling, see ruling
developable surface, 7, 9
direct sum of curves, 43
directrix, see ruling
directrix of a curve, 43
discriminant set, 59
distinguished local basis, 38
drill, 38, 45, 82
dual curve, 58
dual defect, 55, 89, 110, 112, 117
dual projective space, 34
dual variety, 52, 88, 118
Duality Theorem, 33, 55

Euler sequence, 34
Euler's formula, 15

extended Hesse matrix, see Hesse matrix
Extension Lemma, 21, 29

Fano variety, 50
flag variety, 32
flexes, 5
focal hypersurface, 68
focal variety, 68, 90

Gauß almost ruling, 103
Gauß curvature, 3, 5
Gauß defect, 89, 99, 109, 111
Gauß fiber, 87
Gauß image, 87, 88, 103, 121
Gauß map, 1, 85, 124
    fibers of, 101
    singular locus of, 107
Gauß rank, 87, 89
Grassmannian, 22
    tangent space, 35
    Wedge Criterion, 25

Hesse matrix, 5, 6, 96, 97
higher fundamental forms, 127
higher osculating spaces, 127

incidence variety, 49
Irreducibility Theorem, 14
irreducible component, 12
irreducible variety, 12

join, 51, 70
    expected dimension, 52
join correspondence, 122

Linearity Theorem, 3, 87, 101
local basis, 37, 63
local lifting, 9

moving vector, 35, 37

nondegenerate curve, 45, 57
nonordinary variety, 116
Normal Form Lemma, 47
normal space, 94

List of Symbols